D0874208

Longman Mathematical Texts

A short course in
General Relativity

J. Foster
University of Sussex

and

J. D. Nightingale
State University of New York,
College at New Paltz

Longman

London and New York

Longman Group Limited London

Associated companies, branches and representatives throughout the world

Published in the United States of America by Longman Inc., New York

©Longman Group Limited 1979

First published 1979

British Library Cataloguing in Publication Data

Foster, J
  A short course in general relativity. – (Longman
  mathematical texts).
  1. General relativity (Physics)
  I. Title II. Nightingale, J D
  530.1'1 QC173.6 78–40859

  ISBN 0–582–44194–3

Printed in Great Britain by Richard Clay (The Chaucer Press), Ltd, Bungay, Suffolk

Preface

This book provides a short course in general relativity, intended primarily for senior undergraduates or first-year graduate students in physics, mathematics, or related subjects such as astrophysics. Our intention was to produce a book suitable for those who may only take one course in the subject, typically of one or two terms, or one semester duration, but it should also serve as an introduction to the excellent and more comprehensive texts which have appeared in recent years.

Most students approaching the subject require an introduction to tensors, which provide the language of relativity, and these are dealt with in Chapter 1 and the first half of Chapter 2. The latter half of Chapter 2 discusses the geodesic equations, Chapter 3 the field equations, and Chapter 4 applies the results learned to physics in the vicinity of a massive object. Throughout we have tried to compare new results with their Newtonian counterparts. Chapters 5 and 6, on gravitational radiation and cosmology respectively, give further applications of the theory, but students wanting a more detailed knowledge of these topics (and indeed all topics) would have to turn to the longer texts referred to in the body of the book. We finish with an appendix, where special relativity is reviewed, and presented in a form which makes contacts with the general theory easier to establish. Chapters 5 and 6 are independent, and either or both could be omitted to produce a shorter course. Exercises have been provided at the ends of most sections, and problems at the ends of chapters. The former are quite often straightforward (but possibly tedious) verifications needed for a first reading of the book, while the latter are conceivably not so necessary.

General relativity is becoming much more of an experimental subject, and so that the reader may savour something of the way modern technology is brought to bear on the problems which beset experimenters, we have given references (mainly to periodical articles) where appropriate.

Independently, and quite coincidentally, much of the material here has been taught during the past decade at both New Paltz and Sussex, and its consolidation into a single text was the result of one of us taking his sabbatical leave in England. While responsibility for errors is entirely the authors' we would like to mention with gratitude Bob Marchini, John McNamara, John Ray, Eric Shugart and Stacie Swingle, all of whom have been of assistance in one way or another, and not least our wives. Early work on the book owes much to the help and encouragement of Arlene Nightingale, while the final typescript was accurately produced at extremely short notice by Jill Foster. Finally, we would like to thank Professor Alan Jeffrey for his encouragement and the staff at Longman for their courteous cooperation.

Contents

Introduction

The originator of the general theory of relativity was Einstein, and in 1919 he wrote [1]: *The special theory, on which the general theory rests, applies to all physical phenomena with the exception of gravitation; the general theory provides the law of gravitation and its relation to the other forces of nature.* The claim that the general theory provides the law of gravitation does not mean that H. G. Wells' Mr Cavor could now introduce an antigravity material and glide up to the Moon, nor, for example, that we might produce intense permanent gravitational fields in the laboratory, as we can electric fields. It means only that all the properties of gravity of which we are aware are explicable by the theory, and that gravity is essentially a matter of geometry. Before saying how we get to the general from the special theory, we must first discuss the principle of equivalence.

In electrostatics, when a test particle of charge q and inertial mass m_i is placed in a static field $\mathbf{E}$, it experiences a force $q\mathbf{E}$, and undergoes an acceleration

$$\mathbf{a} = (q/m_i)\mathbf{E}. \tag{I.1}$$

In contrast, a test particle of gravitational mass m_g and inertial mass m_i placed in a gravitational field $\mathbf{g}$ experiences a force $m_g\mathbf{g}$, and undergoes an acceleration

$$\mathbf{a} = (m_g/m_i)\mathbf{g}. \tag{I.2}$$

It is an experimental fact (known since Galileo's time) that different particles placed in the same gravitational field acquire the same acceleration (see Fig. I.1(a)). This implies that the ratio m_g/m_i appearing in equation (I.2) is the *same* for all particles, and by an appropriate choice of units this ratio may be taken to be unity. This equivalence of gravitational and inertial mass (which allows us to drop the qualification, and simply refer to *mass*) has been checked experimentally by Eötvös (in 1889 and 1922), and more recently and more accurately (to one part in

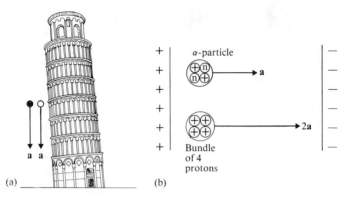

*Fig. I.*1   Test particles in (a) a gravitational field, and (b) an electrostatic field.

10^{11}) by Dicke and his coworkers (in the 1960s). In contrast, the ratio q/m_i occurring in equation (I.1) is not the same for all particles (see Fig. I.1(b)).

With these preliminaries over, we may now consider the principle of equivalence, and it is instructive to do so from the point of view of Einstein's freely falling elevator. If we consider a projectile shot from one side of the elevator cabin to the other, the projectile appears to go in a straight line (the elevator cable being cut) rather than in the usual curved trajectory. Projectiles that are released from rest relative to the cabin remain floating weightless in the cabin. Of course, if the cabin is left to fall for a long time, the particles gradually draw closer together, since they are falling down radial lines towards a common point which is the centre of the Earth. However, if we make the proviso that the cabin is in this state for a short time, as well as being spatially small enough for the neglect of tidal forces in general, then the freely falling cabin (which may have X, Y, Z-coordinates chalked on its walls, as well as a cabin clock measuring time T) looks remarkably like an inertial frame of reference, and therefore the laws of special relativity hold sway inside the cabin. (The cabin must not only occupy a small region of spacetime, it must also be non-rotating with respect to distant matter in the universe [2].) All this follows from the fact that the acceleration of any particle relative to the cabin is zero because they both have the same acceleration relative to the Earth, and we see that the equivalence of inertial and gravitational mass is an essential feature of the discussion. We may incorporate these ideas into the *principle of equivalence*, which is this: *In a freely falling (non-rotating) laboratory occupying a small region of spacetime, the laws of physics are the laws of special relativity* [3].

As a result of the above discussion, the reader should not believe that we can actually transform gravity away by turning to a freely falling reference frame. It is absolutely impossible to transform away a permanent gravitational field of the type associated with a star (as we shall see in Chapter 3), but it is possible to get closer and closer to an ideal inertial reference frame if we make our laboratory occupy smaller and smaller regions of spacetime.

The way in which Einstein generalised the special theory so as to incorporate gravitation was extremely ingenious, and without precedent in the history of science. Gravity was no longer to be regarded as a force, but as a manifestation of the curvature of spacetime itself. The new theory, known as the *general theory of relativity* (or *general relativity* for short), yields the special theory as an approximation in exactly the way the principle of equivalence requires. Because of the curvature of spacetime, it cannot be formulated in terms of coordinate systems based on inertial frames, as the special theory can, and we therefore use arbitrary coordinate systems. Indeed, global inertial frames can no longer be defined, the nearest we can get to them being freely falling non-rotating frames valid in limited regions of spacetime only. A full explanation of what is involved is given in Chapter 2, but we can give a limited preview here.

In special relativity, the invariant expression which defines the proper time τ is given by

$$c^2 \, \mathrm{d}\tau^2 = \eta_{\mu\nu} \, \mathrm{d}X^\mu \, \mathrm{d}X^\nu, \tag{I.3}$$

where the four coordinates X^0, X^1, X^2, X^3 are given in terms of the usual coordinates T, X, Y, Z by

$$X^0 \equiv cT, \qquad X^1 \equiv X, \qquad X^2 \equiv Y, \qquad X^3 \equiv Z. \tag{I.4}$$

(See Section A.0, but note the change to capital letters. See also Section 1.1 for an explanation of the summation convention.) If we change to arbitrary coordinates x^μ, which may be defined in terms of the X^μ in any way whatsoever (they may, for example, be linked to an accelerating or rotating frame), then the expression (I.3) takes the form

$$c^2 \, d\tau^2 = g_{\mu\nu} \, dx^\mu \, dx^\nu, \tag{I.5}$$

where

$$g_{\mu\nu} = \eta_{\rho\sigma} \frac{\partial X^\rho}{\partial x^\mu} \frac{\partial X^\sigma}{\partial x^\nu}.$$

This follows from the fact that $dX^\rho = (\partial X^\rho / \partial x^\mu) \, dx^\mu$. In terms of the coordinates X^μ, the equation of motion of a free particle is

$$d^2 X^\mu / d\tau^2 = 0, \tag{I.6}$$

which, in terms of the arbitrary coordinates, becomes

$$\frac{d^2 x^\mu}{d\tau^2} + \Gamma^\mu_{\nu\sigma} \frac{dx^\nu}{d\tau} \frac{dx^\sigma}{d\tau} = 0, \tag{I.7}$$

where

$$\Gamma^\mu_{\nu\sigma} = \frac{\partial x^\mu}{\partial X^\rho} \frac{\partial^2 X^\rho}{\partial x^\nu \partial x^\sigma},$$

as a short calculation (involving the chain rule) shows. Einstein's proposals for the general theory were that in any coordinate system the proper time should be given by an expression of the form (I.5), and that the equation of motion of a free particle (i.e. one moving under the influence of gravity alone, gravity no longer being a force) should be given by an expression of the form (I.7), but that (in contrast to the spacetime of special relativity) *there are no preferred coordinates X^μ which will reduce these to the forms (I.3) and (I.6)*. This is the essential difference between the spacetimes of special and general relativity. The curvature of spacetime (and therefore gravity) is carried by the $g_{\mu\nu}$, and as we shall see, there is a sense in which these quantities may be regarded as gravitational potentials. We shall also see that the $\Gamma^\mu_{\nu\sigma}$ are determined by the $g_{\mu\nu}$, and that it is always possible to introduce *local* inertial coordinate systems of *limited extent* in which $g_{\mu\nu} \simeq \eta_{\mu\nu}$ and $\Gamma^\mu_{\nu\sigma} \approx 0$, so that equations (I.3) and (I.6) hold as approximations. We thus recover special relativity as an approximation, and in a way which ties in with our discussion of the principle of equivalence.

Because the introduction of curvature forces us to use arbitrary coordinate systems, we need to formulate the theory in a way which is valid in all coordinate systems. This we do by using tensors, the necessary algebra for which is developed in Chapter 1; the way these fit into the theory is explained in Chapter 2. It might be thought that this arbitrariness causes problems, because the coordinates lose the simple physical meanings that the preferred coordinates X^μ of special relativity have. However, we still have contact with the special theory at the local level, and in this way problems of physical meaning and the correct formulation of equations may be overcome. The basic idea is contained in the *principle of general covariance*, which may be stated as follows: *A physical equation of general relativity is generally true in all coordinate systems if (a) the equation is a tensor equation (i.e. it preserves its form under general coordinate transformations), and (b) the equation is true in special relativity.* The way in which this principle works, and the reason why it works, are explained in Section 2.6.

General relativity should not only reduce to special relativity in the appropriate limit, it should also yield Newtonian gravitation as an approximation. Contacts and comparisons with Newtonian theory are made in Sections 2.7, 2.8, 2.9 and 2.10, and extensively in Chapter 4, where we discuss physics in the vicinity of a massive object. These reveal differences between the two theories which provide possible experimental tests of the general theory, and for convenience we list here the experimental and observational evidence concerning these tests, the so-called five tests of general relativity.

1. *Perihelion advance.* General relativity predicts an anomalous advance of the perihelion of planetary orbits. The following (and many more) observations exist for the solar system [4]:

Mercury $43.11 \pm 0.45''$ per century,

Venus $8.4 \pm 4.8''$ per century,

Earth $5.0 \pm 1.2''$ per century.

The predicted values are $43.03''$, $8.6''$ and $3.8''$ respectively.

2. *Deflection of light.* General relativity predicts that light deviates from rectilinear motion near massive objects. The following (and many more) observed deflections exist for light passing the Sun at grazing incidence:

1919   Greenwich Observatory $1.98 \pm 0.16''$,

1922 Lick Observatory $1.82 \pm 0.20''$,

1947 Yerkes Observatory $2.01 \pm 0.27''$,

1972 Mullard Radio Astronomy Observatory, Cambridge
      (using radio sources and interferometers) $1.82 \pm 0.14''$.

The predicted value is $1.75''$.

3. *Spectral shift.*   General relativity predicts that light emanating from near a massive object is red-shifted, while light falling towards a massive object is blue-shifted. Numerous observations of the spectra of white dwarfs, as well as the remarkable terrestrial experiments carried out at the Jefferson Laboratory [5] verify the general-relativistic prediction.

4. *Time delay in radar sounding.*   General relativity predicts a time delay in radar sounding due to the gravitational field of a massive object. Experiments involving the radar sounding of Venus, Mercury and the spacecrafts *Mariner* 6 and 7, performed in the 1960s and 1970s, have yielded agreement with the predicted values to well within the experimental uncertainties [6].

5. *Geodesic effect.*   General relativity predicts that the axis of a gyroscope which is freely orbiting a massive object should precess. For a gyroscope in a near-Earth orbit this precession amounts to $8''$ per year, and an experiment involving a gyroscope in an orbiting satellite is (at the time of writing) being prepared [7].

Finally, let us say something about the notation used in this book. Wherever possible we have chosen it to coincide with that of the more recent and influential texts on general relativity. Greek suffixes (μ, ν, etc.) have the range $0, 1, 2, 3$, while English suffixes from the middle of the alphabet (i, j, etc.) have the range $1, 2, 3$. The signature of the metric tensor is -2, i.e. $\eta_{00} = 1$, $\eta_{11} = \eta_{22} = \eta_{33} = -1$. Rather than use gravitational units in which the gravitational constant G and the speed of light c are unity, we have retained G and c throughout, expect in Chapter 6 where $c = 1$. In the sections on tensor algebra, tensor analysis and curvature, the underlying space or manifold is of arbitrary dimension, and we have used English suffixes from the beginning of the alphabet (a, b, etc.) to denote the arbitrary range $1, 2, \ldots, N$.

Where an equation defines some quantity or operation, the symbol $\equiv$ is used on its first occurrence, and occasionally thereafter as a reminder. Important equations are displayed between parallel lines.

Notes

1. *The Times*, London, 28 November 1919.
2. This statement is related to *Mach's principle*. For a discussion, see Weinberg, 1972, §3.
3. Some authors distinguish between weak and strong equivalence. Our statement is the strong statement; the weak one refers to freely falling particles only, and not to the whole of physics.
4. The figures are taken from Duncombe, 1956.
5. See Pound and Rebka, 1960.
6. See Shapiro, 1968; Shapiro *et al.*, 1971; and Anderson *et al.*, 1975.
7. See the paper by Everitt, Fairbank and Hamilton in Carmeli *et al.*, 1970.

Vectors and tensors

1.0 Introduction

In this first chapter we deal with some algebraic preliminaries,
namely the concepts of vector spaces, their duals, and spaces
which may be derived from these by the process of tensor multiplication. The treatment is quite general, though restricted to real
finite-dimensional vector spaces, and we pay particular attention
to the effect of changes of bases on components, since this aspect
is of particular importance when the ideas developed here are
applied to general relativity in subsequent chapters.

1.1 Vector spaces

We shall not attempt a formal definition of a vector space, but
assume that the reader has some familiarity with the concept. The
excellent text by Halmos is a suitable introduction for those new
to the concept [1].

The essential features of a vector space are that it is a set of
vectors on which are defined two operations, namely addition of
vectors and the multiplication of vectors by *scalars*; that there is
a zero vector in the space; and that each vector in the space has
an inverse such that the sum of a vector and its inverse equals
the zero vector. It may be helpful to picture the set of vectors
comprising a vector space as a set of arrows emanating from
some origin, with addition of vectors given by the usual paral-
lelogram law, and multiplication of a vector by a scalar as a
scaling operation which changes its length but not its direction,
with the proviso that if the scalar is negative, then the scaled
vector will lie in the same line as the original, but point in the
opposite direction. In this picture the zero vector is simply the
point which is the origin (an arrow of zero length), and the inverse of a given vector is one of the same length and in the same
line as the given vector, but pointing in the opposite direction.

In this text our scalars are real numbers, so our vector spaces are termed real. More generally, the scalars could be taken from any field F, giving a vector space over F. We denote the field of real numbers by $\mathbb{R}$, and to distinguish vectors from scalars the former will be printed in bold-faced type ($\mathbf{0}$, $\mathbf{e}$, $\mathbf{v}$, $\boldsymbol{\lambda}$, etc.).

The notion of linear independence is of central importance in vector-space theory. If, for any scalars $\lambda^1, \ldots, \lambda^K$,

$$\lambda^1 \mathbf{v}_1 + \lambda^2 \mathbf{v}_2 + \cdots + \lambda^K \mathbf{v}_K = \mathbf{0} \tag{1.1.1}$$

implies that $\lambda^1 = \lambda^2 = \cdots = \lambda^K = 0$, then the set of vectors $\{\mathbf{v}_1, \mathbf{v}_2, \ldots, \mathbf{v}_K\}$ is said to be *linearly independent*. A set of vectors which is not linearly independent is *linearly dependent*. Thus for a linearly dependent set $\{\mathbf{v}_1, \ldots, \mathbf{v}_K\}$ there exists a non-trivial linear combination of the vectors which equals the zero vector. That is, there exist scalars $\lambda^1, \ldots, \lambda^K$, not all zero (though some may be) such that

$$\lambda^1 \mathbf{v}_1 + \lambda^2 \mathbf{v}_2 + \cdots + \lambda^K \mathbf{v}_K = \mathbf{0}. \tag{1.1.2}$$

It is appropriate here to say something about notation. In the above we have labelled the vectors with a subscript and the scalars with a superscript. At first sight it may be thought that the use of a superscript will lead to confusion with powers, but since we are mainly concerned with linear properties, powers seldom arise; when they do, bracketing removes any ambiguity, e.g. $(\lambda^1)^2$ denotes the square of λ^1. As we shall see, the use of subscripts and superscripts leads to a remarkably efficient notation, the efficiency of which is further improved by *Einstein's summation convention*. This is that if in any expression the same letter appears as a superscript and also as a subscript, then summation over all possible values of the letter is implied. For example, the linear combination

$$\lambda^1 \mathbf{v}_1 + \cdots + \lambda^K \mathbf{v}_K = \sum_{a=1}^{K} \lambda^a \mathbf{v}_a$$

appearing above is written simply as $\lambda^a \mathbf{v}_a$, the range of summation from $a = 1$ to $a = K$ being gleaned from the context. If in any expression we have more than one range of summation, then distinctions may be made by using different alphabets (or different parts of the same alphabet) for different ranges. We shall in general use small English letters a, b, c, etc., to denote a general range from 1 to N, and small Greek letters μ, ν, σ, etc., to denote the range 0, 1, 2, 3 of relativity. We shall make further parenthetical comments on notation as need arises.

A set of vectors which has the property that every vector **v** in a vector space T may be written as a linear combination of its members is said to *span* the space T. Thus the set $\{\mathbf{v}_1, \ldots, \mathbf{v}_K\}$ spans T if every vector $\mathbf{v} \in T$ may be expressed as

$$\mathbf{v} = \lambda^a \mathbf{v}_a, \tag{1.1.3}$$

for some scalars $\lambda^1, \ldots, \lambda^K$. (The symbol $\in$ is read as "belonging to", or as "belongs to", depending on the context.) If a set of vectors both spans T and is linearly independent, then it is a *basis* of T, and we shall restrict ourselves to vector spaces having finite bases. In this case it is possible to show that all bases of a given vector space T contain the same number of members, and this number is called the *dimension* of the space T [2].

Let $\{\mathbf{e}_1, \ldots, \mathbf{e}_N\}$ (or $\{\mathbf{e}_a\}$ for short) be a basis of an N-dimensional vector space T, so any $\boldsymbol{\lambda} \in T$ may be written as $\boldsymbol{\lambda} = \lambda^a \mathbf{e}_a$ for some scalars λ^a. This expression for $\boldsymbol{\lambda}$ in terms of the $\mathbf{e}_a$ is unique, for if $\boldsymbol{\lambda} = \tilde{\lambda}^a \mathbf{e}_a$, then subtraction gives $(\lambda^a - \tilde{\lambda}^a)\mathbf{e}_a = \mathbf{0}$, which implies that $\lambda^a = \tilde{\lambda}^a$, since basis vectors are independent. The scalars λ^a are the *components* of $\boldsymbol{\lambda}$ relative to the basis $\{\mathbf{e}_a\}$.

The last task of this section is to see how the components of a vector transform when a new basis is introduced. Let $\{\mathbf{e}_{a'}\}$ be a new basis of T, and let $\lambda^{a'}$ be the components of $\boldsymbol{\lambda}$ relative to this new basis. So

$$\boldsymbol{\lambda} = \lambda^{a'} \mathbf{e}_{a'}. \tag{1.1.4}$$

(Note that we use the same kernel letter λ for the vector $\boldsymbol{\lambda}$ and its components λ^a or $\lambda^{a'}$, and that the basis to which the components are related is distinguished by the marks, or lack of them, on the superscript. In a similar way the "unprimed" basis $\{\mathbf{e}_a\}$ is distinguished from the "primed" basis $\{\mathbf{e}_{a'}\}$. This notation is a part of the *kernel-index notation* initiated by Schouten and his coworkers [3]. An alternative (and much used) notation involves priming the kernel letter of the components and basis vectors rather than the suffixes, but this has its disadvantages in terms of later economy.) The new basis vectors may each be written as a linear combination of the old:

$$\mathbf{e}_{a'} = X_{a'}^b \mathbf{e}_b, \tag{1.1.5}$$

and conversely the old as a linear combination of the new:

$$\mathbf{e}_c = X_c^{a'} \mathbf{e}_{a'}. \tag{1.1.6}$$

(Although we use the same kernel letter X, the N^2 numbers $X_{a'}^b$ are different from the N^2 numbers $X_b^{c'}$, the positions of the primes indicating the difference.) Substitution for $\mathbf{e}_{a'}$ from equation (1.1.5) in (1.1.6) yields

$$\mathbf{e}_c = X_c^{a'} X_{a'}^b \mathbf{e}_b. \tag{1.1.7}$$

By the uniqueness of components we then have

$$X_c^{a'} X_{a'}^b = \delta_c^b, \tag{1.1.8}$$

where δ_c^b is the *Kronecker delta* defined by the statement that $\delta_c^b = 0$ if $b \neq c$, but $\delta_c^b = 1$ if $b = c$. (Note that we cannot say $\delta_b^b = 1$, for b then appears both as a superscript and a subscript, and according to our convention summation is implied. In fact $\delta_b^b = N$, the dimension of T.) Similarly, by substituting for $\mathbf{e}_b$ in equation (1.1.5) from (1.1.6) and changing the lettering of suffixes, we also deduce that

$$X_{a'}^b X_b^{c'} = \delta_a^c. \tag{1.1.9}$$

(Some users of this notation prefer to write the right-hand side of equation (1.1.9) as $\delta_{a'}^{c'}$ to give a more balanced appearance.)

Substitution for $\mathbf{e}_{a'}$ from equation (1.1.5) in (1.1.4) yields

$$\boldsymbol{\lambda} = \lambda^{a'} X_{a'}^b \mathbf{e}_b, \tag{1.1.10}$$

and by the uniqueness of components,

$$\lambda^b = X_{a'}^b \lambda^{a'}. \tag{1.1.11}$$

Then

$$X_a^{c'} \lambda^a = X_a^{c'} X_{b'}^a \lambda^{b'} = \delta_b^c \lambda^{b'} = \lambda^{c'}, \tag{1.1.12}$$

on changing the lettering of suffixes. (This change of lettering was simply to avoid a letter appearing more than twice, which would make a nonsense of the notation. Repeated suffixes which imply summation are called *dummy suffixes*, and they may be replaced by any letter.)

To recap, if the primed and unprimed bases are related by

$$\mathbf{e}_{a'} = X_{a'}^b \mathbf{e}_b, \qquad \mathbf{e}_a = X_a^{b'} \mathbf{e}_{b'}, \tag{1.1.13}$$

then the components are related by

$$\lambda^{a'} = X_b^{a'}\lambda^b, \qquad \lambda^a = X_{b'}^a\lambda^{b'}, \tag{1.1.14}$$

and

$$X_{b'}^a X_c^{b'} = \delta_c^a, \qquad X_b^{a'} X_{c'}^b = \delta_c^a. \tag{1.1.15}$$

In the above we have made parenthetical comments on the notation employed, principally because it is unlike that used elsewhere in physics. Indeed, the reader familiar with matrix algebra may be wondering why we have not used the notations of that subject, for equations (1.1.14) may readily be translated into relations between column vectors, with either λ^a or $\lambda^{a'}$ as their ath entries, and square matrices, with either $X_{b'}^a$ or $X_b^{a'}$ as the entry in the ath row and bth column. Equations (1.1.15) would then express the fact that the product of the two matrices involved is the unit matrix, i.e. that each is the inverse of the other. Suffice it to say that our notation has advantages of efficiency and economy, and a certain aesthetic appeal. Moreover, the language of matrices is incapable of dealing with tensors except in an extremely cumbersome way. We shall, however, make some concession to matrices, and refer to the $X_{b'}^a$ and $X_b^{a'}$ occurring in equations (1.1.13–15) as *matrix elements*. The notation $[A_{ab}]$ will be used to denote the matrix having A_{ab} as the entry in its ath row and bth column; similarly $[A_b^a]$ will be used to denote the matrix having A_b^a as the entry in the ath row and bth column, so the superscript indicates the row and the subscript the column. The notation $|A_{ab}|$ and $|A_b^a|$ will be used for determinants.

Exercises 1.1

1. Show that:
   (a) $\lambda^a\delta_a^b = \lambda^b$,
   (b) $\lambda^a\delta_a^b\mu_b = \lambda^a\mu_a = \lambda^b\mu_b$.

2. Derive the result (1.1.9).

1.2 Dual spaces

Although we have suggested that it may be helpful to visualise the vectors of a vector space as a set of arrows emanating from an origin, in some ways this picture can be misleading, for many sets of objects bearing no resemblance to arrows constitute vector spaces under suitable definitions of addition and scalar multiplication. Among such objects are functions.

Let us confine our attention to real-valued functions defined on a real vector space T. In mathematical language such a function f would be written as $f: T \to \mathbb{R}$, indicating that it maps vectors of T into real numbers. The set of all such functions may be given a vector-space structure by defining:

(a) the sum $f + g$ of two functions f and g by

   $(f + g)(\mathbf{v}) = f(\mathbf{v}) + g(\mathbf{v})$ for all $\mathbf{v} \in T$;

(b) the product αf of the scalar α and the function f by

   $(\alpha f)(\mathbf{v}) = \alpha(f(\mathbf{v}))$ for all $\mathbf{v} \in T$;

(c) the zero function 0 by

   $0(\mathbf{v}) = 0$ for all $\mathbf{v} \in T$

   (where on the left 0 is a function, while on the right it is the real number zero, there being no particular advantage in using different symbols);

(d) the inverse $-f$ of the function f by

   $(-f)(\mathbf{v}) = -(f(\mathbf{v}))$ for all $\mathbf{v} \in T$.

That this does indeed define a vector space may be verified by checking the vector-space axioms given in Halmos [1].

The space of all real-valued functions on a vector space T is too large for our purpose, and we shall restrict ourselves to those functions which are *linear*. That is, those functions f which satisfy

$$f(\alpha \mathbf{u} + \beta \mathbf{v}) = \alpha f(\mathbf{u}) + \beta f(\mathbf{v}), \tag{1.2.1}$$

for all α, $\beta \in \mathbb{R}$ and all $\mathbf{u}$, $\mathbf{v} \in T$. Real-valued linear functions on a real vector space are usually called *linear functionals*. It is a simple matter to check that the sum of two linear functionals is itself a linear functional, and that multiplication of a linear functional by a scalar yields a linear functional. These observations are sufficient to show that the set of linear functionals on a vector space T is itself a vector space. This space is the *dual* of T, and we denote it by T^*.

Since linear functionals are vectors we shall from here on use bold-faced type for them. So if $\boldsymbol{\lambda} \in T$ and $\boldsymbol{\mu} \in T^*$, then $\boldsymbol{\mu}(\boldsymbol{\lambda}) \in \mathbb{R}$, i.e. is a scalar (despite the bold-faced type).

We now have two types of vectors, those in T and those in T^*. To distinguish them, those in T are called *contravariant vectors*, while those in T^* are called *covariant vectors*. As a further distinguishing feature, basis vectors of T^* will carry superscripts, and components of vectors in T^* will carry subscripts. Thus if $\{\mathbf{e}^a\}$ is a basis of T^*, then $\boldsymbol{\lambda} \in T^*$ has a unique expression $\boldsymbol{\lambda} = \lambda_a \mathbf{e}^a$ in terms of components.

The use of the lower-case letter a in the implied summation above suggests that the range of summation is from 1 to N, the dimension of T, i.e. that T^* has the same dimension as T. This is in fact the case, as we shall now prove by showing that a given basis $\{\mathbf{e}_a\}$ of T induces in a natural way a *dual basis* $\{\mathbf{e}^a\}$ of T^* having N members satisfying $\mathbf{e}^a(\mathbf{e}_b) = \delta_b^a$.

We start by defining $\mathbf{e}^a$ to be the real-valued function which maps any vector $\boldsymbol{\lambda} \in T$ into the real number which is its ath component λ^a relative to $\{\mathbf{e}_a\}$, i.e. $\mathbf{e}^a(\boldsymbol{\lambda}) = \lambda^a$, for all $\boldsymbol{\lambda} \in T$. This gives us N real-valued functions which clearly satisfy $\mathbf{e}^a(\mathbf{e}_b) = \delta_b^a$, and it remains to show that they are linear, and that they constitute a basis of T^*. The former is readily checked. As for the latter we proceed as follows.

For any $\boldsymbol{\mu} \in T^*$ we can define N real numbers μ_a by $\boldsymbol{\mu}(\mathbf{e}_a) \equiv \mu_a$. *Then for any* $\boldsymbol{\lambda} \in T$,

$$\boldsymbol{\mu}(\boldsymbol{\lambda}) = \boldsymbol{\mu}(\lambda^a \mathbf{e}_a) = \lambda^a \boldsymbol{\mu}(\mathbf{e}_a) \qquad \text{(by the linearity of } \boldsymbol{\mu}\text{)}$$

$$= \lambda^a \mu_a = \mu_a \mathbf{e}^a(\boldsymbol{\lambda}).$$

Thus for any $\boldsymbol{\mu} \in T^*$ we have $\boldsymbol{\mu} = \mu_a \mathbf{e}^a$, showing that $\{\mathbf{e}^a\}$ spans T^*, and there remains the question of the independence of $\{\mathbf{e}^a\}$. This is answered by noting that a relation $x_a \mathbf{e}^a = \mathbf{0}$, where $x_a \in \mathbb{R}$ and $\mathbf{0}$ is the zero functional, implies that

$$0 = x_a \mathbf{e}^a(\mathbf{e}_b) = x_a \delta_b^a = x_b \qquad \text{for all } b.$$

From the above it may be seen that given a basis $\{\mathbf{e}_a\}$ of T, the components μ_a of $\boldsymbol{\mu} \in T^*$ relative to the dual basis $\{\mathbf{e}^a\}$ are given by $\mu_a = \boldsymbol{\mu}(\mathbf{e}_a)$.

A change of basis (1.1.13) in T induces a change of the dual basis. Let us denote the dual of the primed basis $\{\mathbf{e}_{a'}\}$ by $\{\mathbf{e}^{a'}\}$, so by definition $\mathbf{e}^{a'}(\mathbf{e}_{b'}) = \delta^a_b$, and $\mathbf{e}^{a'} = Y^{a'}_b \mathbf{e}^b$ for some $Y^{a'}_b$. Then

$$
\begin{aligned}
\delta^a_b = \mathbf{e}^{a'}(\mathbf{e}_{b'}) &= Y^{a'}_d \mathbf{e}^d (X^c_{b'} \mathbf{e}_c) \\
&= Y^{a'}_d X^c_{b'} \mathbf{e}^d (\mathbf{e}_c) \qquad \text{(by the linearity of the } \mathbf{e}^d) \\
&= Y^{a'}_d X^c_{b'} \delta^d_c = Y^{a'}_c X^c_{b'}.
\end{aligned}
$$

Multiplying by $X^{b'}_d$ gives $X^{a'}_d = Y^{a'}_d$. Thus under a change of basis of T given by equation (1.1.13), the dual bases of T^* transform according to

$$
\mathbf{e}^{a'} = X^{a'}_b \mathbf{e}^b, \qquad \mathbf{e}^a = X^a_{b'} \mathbf{e}^{b'}. \tag{1.2.2}
$$

It is readily shown that the components of $\boldsymbol{\mu} \in T^*$ relative to the dual bases transform according to

$$
\mu_{a'} = X^b_{a'} \mu_b, \qquad \mu_a = X^{b'}_a \mu_{b'}. \tag{1.2.3}
$$

So the same matrix $[X^{a'}_b]$ and its inverse $[X^a_{b'}]$ are involved, but their roles relative to basis vectors and components are interchanged.

Given T and a basis $\{\mathbf{e}_a\}$ of it, we have seen how to construct its dual T^* with dual basis $\{\mathbf{e}^a\}$ satisfying $\mathbf{e}^a(\mathbf{e}_b) = \delta^a_b$. We can apply this process again to arrive at the dual T^{**} of T^*, with dual basis $\{\mathbf{f}_a\}$ say, satisfying $\mathbf{f}_a(\mathbf{e}^b) = \delta^b_a$, and vectors $\boldsymbol{\lambda} \in T^{**}$ may be expressed in terms of components as $\boldsymbol{\lambda} = \lambda^a \mathbf{f}_a$. Under a change of basis of T, components of vectors in T transform according to $\lambda^{a'} = X^{a'}_b \lambda^b$. This induces a change of dual basis of T^*, under which components of vectors in T^* transform according to $\mu_{a'} = X^b_{a'} \mu_b$. In turn, this induces a change of basis of T^{**}, under which it is readily seen that components of vectors in T^{**} transform according to $\lambda^{a'} = X^{a'}_b \lambda^b$ (because the inverse of the inverse of a matrix is the matrix itself). That is, the components of vectors in T^{**} transform in exactly the same way as the components of vectors in T. This means that if we set up a one-to-one correspondence between vectors in T and T^{**} by making $\lambda^a \mathbf{e}_a$ in T correspond to $\lambda^a \mathbf{f}_a$ in T^{**}, where $\{\mathbf{f}_a\}$ is the dual of the dual of $\{\mathbf{e}_a\}$, then this correspondence is *basis-independent*. A basis-independent one-to-one correspondence between vector spaces is called a *natural isomorphism*, and naturally isomorphic vector spaces are usually identified, by identifying corresponding vectors. Consequently we shall identify T^{**} with T.

Exercises 1.2

1. Check that the sum of two linear functionals is itself a linear functional, and that multiplication of a linear functional by a scalar yields a linear functional.
2. Verify that the components of $\boldsymbol{\mu} \in T^*$ relative to the dual bases transform according to equations (1.2.3), as asserted.
3. Identifying T^{**} with T means that a contravariant vector $\boldsymbol{\lambda}$ acts as a linear functional on covariant vectors $\boldsymbol{\mu}$. Show that in terms of components, $\boldsymbol{\lambda}(\boldsymbol{\mu}) = \lambda^a \mu_a$.

1.3 Tensor products

Given a vector space T we have seen how to create a new vector space, namely its dual T^*, but here the process stops (on identifying T^{**} with T). However, it is possible to generate a new vector space from two vector spaces by forming what is called their tensor product. As a preliminary to this we need to define bilinear functionals on a pair of vector spaces.

Let T and U be two real finite-dimensional vector spaces. The *cartesian product* $T \times U$ is the set of all ordered pairs of the form $(\mathbf{v}, \mathbf{w})$, where $\mathbf{v} \in T$ and $\mathbf{w} \in U$. A *bilinear functional* f on $T \times U$ is a real-valued function $f: T \times U \to \mathbb{R}$, which is bilinear, i.e. satisfies

$$f(\alpha\mathbf{u} + \beta\mathbf{v}, \mathbf{w}) = \alpha f(\mathbf{u}, \mathbf{w}) + \beta f(\mathbf{v}, \mathbf{w}),$$

$$\text{for all} \quad \alpha, \beta \in \mathbb{R}, \quad \mathbf{u}, \mathbf{v} \in T \quad \text{and} \quad \mathbf{w} \in U,$$

and

$$f(\mathbf{v}, \gamma\mathbf{w} + \delta\mathbf{x}) = \gamma f(\mathbf{v}, \mathbf{w}) + \delta f(\mathbf{v}, \mathbf{x}),$$

$$\text{for all} \quad \gamma, \delta \in \mathbb{R}, \quad \mathbf{v} \in T \quad \text{and} \quad \mathbf{w}, \mathbf{x} \in U.$$

With definitions of addition, scalar multiplication, the zero function and inverses analogous to those given for linear functionals in Section 1.2, it is a straightforward matter to show that the set of bilinear functionals on $T \times U$ is a vector space, and from here on we shall use bold-faced type for bilinear functionals.

We are now in a position to define the *tensor product* $T \otimes U$ of T and U as the vector space of all bilinear functionals on $T^* \times U^*$. Note that in this definition we use the dual spaces T^* and U^*, and not T and U themselves.

The question naturally arises as to the dimension of $T \otimes U$. It is in fact NM, where N and M are the dimensions of T and U

respectively, and we prove this by showing that from given bases of T and U we can define NM members of $T \otimes U$ which constitute a basis for it.

Let $\{\mathbf{e}^a\}$, $a = 1, \ldots, N$, and $\{\mathbf{f}^\alpha\}$, $\alpha = 1, \ldots, M$, be bases of T^* and U^*, dual to bases $\{\mathbf{e}_a\}$ and $\{\mathbf{f}_\alpha\}$ of T and U respectively. (Note that we use different alphabets for suffixes having different ranges.) Define NM functions $\mathbf{e}_{a\alpha}: T^* \times U^* \to \mathbb{R}$ by

$$\mathbf{e}_{a\alpha}(\boldsymbol{\lambda}, \boldsymbol{\mu}) = \lambda_a \mu_\alpha, \tag{1.3.1}$$

where λ_a are the components of $\boldsymbol{\lambda} \in T^*$ relative to $\{\mathbf{e}^a\}$ and μ_α are those of $\boldsymbol{\mu} \in U^*$ relative to $\{\mathbf{f}^\alpha\}$. In particular

$$\mathbf{e}_{a\alpha}(\mathbf{e}^b, \mathbf{f}^\beta) = \delta_a^b \delta_\alpha^\beta. \tag{1.3.2}$$

It is a simple matter to show that the $\mathbf{e}_{a\alpha}$ are bilinear and so belong to $T \otimes U$. To show that they constitute a basis we must show that they span $T \otimes U$ and that they are independent.

For any $\boldsymbol{\tau} \in T \otimes U$, define NM real numbers $\tau^{a\alpha}$ by $\boldsymbol{\tau}(\mathbf{e}^a, \mathbf{f}^\alpha) \equiv \tau^{a\alpha}$. Then for any $\boldsymbol{\lambda} \in T^*$ and $\boldsymbol{\mu} \in U^*$ we have

$$\begin{aligned}
\boldsymbol{\tau}(\boldsymbol{\lambda}, \boldsymbol{\mu}) &= \boldsymbol{\tau}(\lambda_a \mathbf{e}^a, \mu_\alpha \mathbf{f}^\alpha) \\
&= \lambda_a \mu_\alpha \boldsymbol{\tau}(\mathbf{e}^a, \mathbf{f}^\alpha) \qquad \text{(on using the bilinearity of } \boldsymbol{\tau}\text{)} \\
&= \tau^{a\alpha} \lambda_a \mu_\alpha = \tau^{a\alpha} \mathbf{e}_{a\alpha}(\boldsymbol{\lambda}, \boldsymbol{\mu}).
\end{aligned}$$

So for any $\boldsymbol{\tau} \in T \otimes U$, we have $\boldsymbol{\tau} = \tau^{a\alpha} \mathbf{e}_{a\alpha}$, showing that the set $\{\mathbf{e}_{a\alpha}\}$ spans $T \otimes U$. Moreover, $\{\mathbf{e}_{a\alpha}\}$ is an independent set, for if $x^{a\alpha} \mathbf{e}_{a\alpha} = \mathbf{0}$, then

$$0 = x^{a\alpha} \mathbf{e}_{a\alpha}(\mathbf{e}^b, \mathbf{f}^\beta) = x^{a\alpha} \delta_a^b \delta_\alpha^\beta = x^{b\beta}$$

for all b, β, on using equation (1.3.2).

Thus we have shown that the dimension of $T \otimes U$ is the product of the dimensions of T and U, and that in a natural way bases $\{\mathbf{e}_a\}$ of T and $\{\mathbf{f}_\alpha\}$ of U induce a basis $\{\mathbf{e}_{a\alpha}\}$ of $T \otimes U$, the components $\tau^{a\alpha}$ of any $\boldsymbol{\tau} \in T \otimes U$ relative to this basis being given in terms of the dual bases of T^* and U^* by $\tau^{a\alpha} = \boldsymbol{\tau}(\mathbf{e}^a, \mathbf{f}^\alpha)$.

Let us now investigate how the components $\tau^{a\alpha}$ and the induced basis vectors $\mathbf{e}_{a\alpha}$ transform when new bases are introduced into T and U. Suppose the bases of T and U are changed according to

$$\mathbf{e}_{a'} = X_{a'}^b \mathbf{e}_b, \quad \mathbf{f}_{\alpha'} = Y_{\alpha'}^\beta \mathbf{f}_\beta. \tag{1.3.3}$$

This induces a new basis $\{\mathbf{e}_{a'\alpha'}\}$ of $T \otimes U$, and for any $(\boldsymbol{\lambda}, \boldsymbol{\mu}) \in T^* \times U^*$,

$$\mathbf{e}_{a'\alpha'}(\boldsymbol{\lambda}, \boldsymbol{\mu}) = \lambda_{a'}\mu_{\alpha'} = X_{a'}^b Y_{\alpha'}^\beta \lambda_b \mu_\beta$$
$$= X_{a'}^b Y_{\alpha'}^\beta \mathbf{e}_{b\beta}(\boldsymbol{\lambda}, \boldsymbol{\mu}).$$

So

$$\mathbf{e}_{a'\alpha'} = X_{a'}^b Y_{\alpha'}^\beta \mathbf{e}_{b\beta}. \tag{1.3.4}$$

Similarly, for components (see Exercise 1.3.2),

$$\tau^{a'\alpha'} = X_b^{a'} Y_\beta^{\alpha'} \tau^{b\beta}. \tag{1.3.5}$$

A vector which is a member of the tensor product of two spaces (or more, see below) is called a *tensor*. The tensor product as defined above is a product of spaces. It is possible to define a tensor which is the tensor product $\boldsymbol{\lambda} \otimes \boldsymbol{\mu}$ of individual vectors $\boldsymbol{\lambda} \in T$ and $\boldsymbol{\mu} \in U$ by setting

$$\boldsymbol{\lambda} \otimes \boldsymbol{\mu} \equiv \lambda^a \mu^\alpha \mathbf{e}_{a\alpha}, \tag{1.3.6}$$

where λ^a and μ^α are the components of $\boldsymbol{\lambda}$ and $\boldsymbol{\mu}$ relative to bases of T and U which induce the basis $\{\mathbf{e}_{a\alpha}\}$ of $T \otimes U$. Although this definition is given via bases, it is in fact basis-independent (see Exercise 1.3.3). In particular we have

$$\mathbf{e}_a \otimes \mathbf{f}_\alpha = \boldsymbol{e}_{a\alpha}. \tag{1.3.7}$$

The tensor product $\boldsymbol{\lambda} \otimes \boldsymbol{\mu}$ belongs to $T \otimes U$, but not all tensors in $T \otimes U$ are of this form. Those that are are called *decomposable*.

Having established the basic idea of the tensor product of vector spaces we can extend it to three or more spaces. However, given three spaces T, U and V we can form their tensor product in two ways: $(T \otimes U) \otimes V$ or $T \otimes (U \otimes V)$. These two spaces clearly have the same dimension, and are in fact naturally isomorphic, in the sense that we can set up a basis-independent one-to-one correspondence between their members, just as we did with T and T^{**}. This is done by choosing bases $\{\mathbf{e}^a\}$, $\{\mathbf{f}^\alpha\}$, $\{\mathbf{g}^A\}$ in T, U, V respectively (three ranges, so three alphabets), letting $\tau^{a\alpha A}\mathbf{e}_a \otimes (\mathbf{f}_\alpha \otimes \mathbf{g}_A)$ in $T \otimes (U \otimes V)$ correspond to $\tau^{a\alpha A}(\mathbf{e}_a \otimes \mathbf{f}_\alpha) \otimes \mathbf{g}_A$, and then showing that this correspondence is basis-independent. Because of the natural isomorphism one identifies these spaces, and the notation $T \otimes U \otimes V$ is unambiguous.

An alternative way of defining $T \otimes U \otimes V$ is as the space of trilinear functionals on $T^* \times U^* \times V^*$. This leads to a space which is naturally isomorphic to those of the preceding paragraph, and all three are identified. Other natural isomorphisms exist, for example between $T \otimes U$ and $U \otimes T$, or between $(T \otimes U)^*$ and $T^* \otimes U^*$, and whenever they exist, the spaces are identified.

Exercises 1.3

1. Show that the functions $\mathbf{e}_{a\alpha} : T^* \times U^* \to \mathbb{R}$, defined by equation (1.3.1), are bilinear functionals.
2. Verify the transformation formula for components (equation (1.3.5)).
3. Prove that the definition of the tensor product $\boldsymbol{\lambda} \otimes \boldsymbol{\mu}$ of two vectors $\boldsymbol{\lambda}$ and $\boldsymbol{\mu}$ is basis-independent.

1.4   The space T_s^r

From now on we shall confine our attention to tensor-product spaces obtained by taking repeated tensor products of just one space T and/or its dual T^*. We introduce the following notation:

$$\underbrace{T \otimes T \otimes \cdots \otimes T}_{r \text{ times}} \equiv T^r,$$

$$\underbrace{T^* \otimes T^* \otimes \cdots \otimes T^*}_{s \text{ times}} \equiv T_s,$$

$$T^r \otimes T_s \equiv T_s^r.$$

In particular $T = T^1$ and $T^* = T_1$.

A member of T^r is a *contravariant tensor of order r*, a member of T_s is a *covariant tensor of order s*, while a member of T_s^r is a *mixed tensor of order* $(r + s)$. A member of T^r is also referred to as a *tensor of type* $(r, 0)$, a member of T_s as a *tensor of type* $(0, s)$, and a member of T_s^r as a *tensor of type* (r, s). Note that this nomenclature labels contravariant and covariant vectors as tensors of types $(1, 0)$ and $(0, 1)$ respectively. Scalars may be included in the general scheme of things by regarding them as type $(0, 0)$ tensors.

A basis $\{\mathbf{e}_a\}$ of T (of dimension N) induces a dual basis $\{\mathbf{e}^a\}$ of T^*, and these together yield a basis $\{\mathbf{e}_{a_1 \ldots a_r}^{b_1 \ldots b_s}\}$ of T_s^r. Each tensor $\boldsymbol{\tau} \in T_s^r$ has N^{r+s} unique components relative to this induced basis:

$$\boldsymbol{\tau} = \tau_{b_1 \ldots b_s}^{a_1 \ldots a_r} \, \mathbf{e}_{a_1 \ldots a_r}^{b_1 \ldots b_s}. \tag{1.4.1}$$

A change of basis of T induces a change of basis of T_s^r under which the components transform according to

$$\tau_{b_1'\ldots b_s'}^{a_1'\ldots a_r'} = X_{c_1}^{a_1'}\cdots X_{c_r}^{a_r'} X_{b_1'}^{d_1}\cdots X_{b_s'}^{d_s}\tau_{d_1\ldots d_s}^{c_1\ldots c_r}, \tag{1.4.2}$$

where $X_c^{a'}$ and $X_{b'}^d$ are the matrix elements involved in the change of basis of T. For example, for $\boldsymbol{\tau} \in T_2^1$,

$$\tau_{b'c'}^{a'} = X_d^{a'} X_{b'}^e X_{c'}^f \tau_{ef}^d.$$

Some authors define tensors as objects having components which transform according to equation (1.4.2). This way of looking at tensors may be justified by noting that if with each basis of T there are associated N^{r+s} real numbers, which under a change of basis given by equation (1.1.13) transform according to equation (1.4.2), then these numbers are the components of a tensor $\boldsymbol{\tau}$ of type (r, s). We simply put

$$\boldsymbol{\tau} = \tau_{b_1\ldots b_s}^{a_1\ldots a_r}\, \mathbf{e}_{a_1\ldots a_r}^{b_1\ldots b_s}.$$

A useful theorem for determining tensor character is the so-called *quotient theorem*. Rather than give a general statement of the theorem and its proof, which tend to be obscured by a mass of suffixes, we shall give an example which illustrates the gist of the theorem.

Suppose that with each basis of T there are associated N^3 numbers τ_{bc}^a and that it is known that for *arbitrary* contravariant vectors $\boldsymbol{\lambda}$ with components λ^a, the N^2 numbers $\tau_{bc}^a \lambda^c$ transform as a type $(1, 1)$ tensor under a change of basis. That is

$$\tau_{b'c'}^{a'} \lambda^{c'} = X_d^{a'} X_{b'}^e \tau_{ef}^d \lambda^f, \tag{1.4.3}$$

where $\tau_{b'c'}^{a'}$ are the N^3 numbers associated with the primed basis. Then we may deduce that the τ_{bc}^a are components of a type $(1, 2)$ tensor. For $\lambda^{c'} = X_f^{c'} \lambda^f$, so equation (1.4.3) yields

$$(\tau_{b'c'}^{a'} X_f^{c'} - X_d^{a'} X_{b'}^e \tau_{ef}^d)\lambda^f = 0, \tag{1.4.4}$$

and this holds for arbitrary $\boldsymbol{\lambda}$. Now let $\boldsymbol{\lambda}$ be the *standard vector* having one as its gth component and the others zero, i.e. $\lambda^f = \delta_g^f$. Equation (1.4.4) then gives

$$\tau_{b'c'}^{a'} X_g^{c'} = X_d^{a'} X_{b'}^e \tau_{eg}^d,$$

valid for all g. Multiplying by $X^g_{h'}$ and using $X^{c'}_g X^g_{h'} = \delta^c_h$ gives

$$\tau^{a'}_{b'h'} = X^{a'}_d X^e_{b'} X^g_{h'} \tau^d_{eg},$$

which establishes that the τ^a_{bc} are indeed the components of a type $(1, 2)$ tensor.

This example illustrates the gist of the quotient theorem, which is that if numbers which are candidates for tensor components display tensor character when some of their suffixes are "killed off" by summation with the components of *arbitrary* vectors (or tensors), then this is sufficient to establish the original numbers as the components of a tensor. We shall have occasions to use this theorem later in the book.

Although we are not using matrices to develop the general theory of tensors, they can have their uses for calculating components in particular cases, as this example shows.

Example

Suppose that the dimension of T is 3, and that relative to $\{\mathbf{e}_a\}$ the type $(2, 0)$ tensor $\boldsymbol{\tau}$ has components

$$[\tau^{ab}] \equiv \begin{bmatrix} 1 & 0 & 1 \\ 0 & 1 & 0 \\ 1 & 0 & 1 \end{bmatrix}$$

If $\mathbf{e}_{b'} = X^a_{b'} \mathbf{e}_a$, where

$$[X^a_{b'}] \equiv \begin{bmatrix} 1 & 1 & 1 \\ 0 & 1 & 1 \\ 0 & 0 & 1 \end{bmatrix}, \tag{1.4.5}$$

how do we find the components $\tau^{a'b'}$ of $\boldsymbol{\tau}$ relative to $\{\mathbf{e}_{a'}\}$? Formula (1.4.2) gives

$$\tau^{a'b'} = X^{a'}_c X^{b'}_d \tau^{cd},$$

or, on changing the order on the right-hand side

$$\tau^{a'b'} = X^{a'}_c \tau^{cd} X^{b'}_d. \tag{1.4.6}$$

For matrices such as $[X^a_{b'}]$ and $[X^{a'}_b]$ our convention is to let the superscript label the row and the subscript the column (see Section 1.1), so the matrix version of equation (1.4.6) has to be

$$T' = XTX^\mathsf{T},$$

where $T' = [\tau^{a'b'}]$, $T = [\tau^{ab}]$, $X = [X_b^{a'}]$, and X^T denotes the transpose of X. Now $[X_b^{a'}]$ is the inverse of the displayed matrix (1.4.5), so

$$X = \begin{bmatrix} 1 & -1 & 0 \\ 0 & 1 & -1 \\ 0 & 0 & 1 \end{bmatrix},$$

and

$$\begin{aligned} T' &= \begin{bmatrix} 1 & -1 & 0 \\ 0 & 1 & -1 \\ 0 & 0 & 1 \end{bmatrix} \begin{bmatrix} 1 & 0 & 1 \\ 0 & 1 & 0 \\ 1 & 0 & 1 \end{bmatrix} \begin{bmatrix} 1 & 0 & 0 \\ -1 & 1 & 0 \\ 0 & -1 & 1 \end{bmatrix} \\ &= \begin{bmatrix} 1 & -1 & 0 \\ 0 & 1 & -1 \\ 0 & 0 & 1 \end{bmatrix} \begin{bmatrix} 1 & -1 & 1 \\ -1 & 1 & 0 \\ 1 & -1 & 1 \end{bmatrix} \\ &= \begin{bmatrix} 2 & -2 & 1 \\ -2 & 2 & -1 \\ 1 & -1 & 1 \end{bmatrix}. \end{aligned}$$

The components $\tau^{a'b'}$ can then be picked out from

$$[\tau^{a'b'}] = \begin{bmatrix} 2 & -2 & 1 \\ -2 & 2 & -1 \\ 1 & -1 & 1 \end{bmatrix}.$$

1.5 Combining tensors

So far we have three basic operations with tensors, namely addition of tensors of the same type, multiplication of a tensor by a scalar, and the formation of the tensor product. There is a fourth basic operation with tensors which is most readily explained in terms of components. It is an operation which may be applied to any object characterised by sets of numbers specified by letters carrying superscripts and subscripts, but it takes on a special significance when the numbers are tensor components. This operation is *contraction*, which associates N^{r+s-2} numbers $R_{b_1 \ldots b_{q-1} b_{q+1} \ldots b_s}^{a_1 \ldots a_{p-1} a_{p+1} \ldots a_r}$ with N^{r+s} numbers $Q_{b_1 \ldots b_s}^{a_1 \ldots a_r}$, defined by

$$R_{b_1 \ldots b_{q-1} b_{q+1} \ldots b_s}^{a_1 \ldots a_{p-1} a_{p+1} \ldots a_r} \equiv Q_{b_1 \ldots b_{q-1} c b_{q+1} \ldots b_s}^{a_1 \ldots a_{p-1} c a_{p+1} \ldots a_r}. \tag{1.5.1}$$

That is, by putting a subscript equal to a superscript and summing, as the summation convention implies. It is clear that there are rs ways that this may be done, each of which leads to a *contraction* of the original sets of numbers.

The special significance that this operation has for tensors is that if the original numbers are the components of a type (r, s) tensor, then their contractions are components of a type $(r-1, s-1)$ tensor. The proof in the general case is somewhat cumbersome, but an example gives the gist of it.

Suppose τ_c^{ab} are the components of a type $(2, 1)$ tensor, and we form $\sigma^a = \tau_b^{ab}$ by contraction. Relative to another (primed) basis this would yield $\sigma^{a'} = \tau_{b'}^{a'b'}$. Then

$$\sigma^{a'} = \tau_{b'}^{a'b'} = \tau_e^{cd} X_c^{a'} X_d^{b'} X_{b'}^e$$
$$= \tau_e^{cd} X_c^{a'} \delta_d^e = \tau_d^{cd} X_c^{a'} = \sigma^c X_c^{a'},$$

showing that the numbers σ^a obtained by contraction are the components of a type $(1, 0)$ tensor (a contravariant vector).

Contraction may be combined with tensor multiplication. For example, if $\boldsymbol{\rho}$ is a type $(1, 1)$ tensor and $\boldsymbol{\sigma}$ a contravariant vector, then the contravariant vector $\boldsymbol{\tau}$ with components $\tau^a = \rho_b^a \sigma^b$ is obtained by *contracting* $\boldsymbol{\rho}$ with $\boldsymbol{\sigma}$. The tensor $\boldsymbol{\tau}$ is a contraction of the type $(2, 1)$ tensor $\boldsymbol{\rho} \otimes \boldsymbol{\sigma}$ with components $\rho_b^a \sigma^c$.

1.6   Special tensors

Consider the tensor $\boldsymbol{\kappa}$ of type $(1, 1)$ defined by specifying its components relative to a given basis $\{\mathbf{e}_a\}$ to be the Kronecker delta. i.e. $\kappa_b^a \equiv \delta_b^a$. Relative to another basis $\{\mathbf{e}_{a'}\}$ its components are

$$\kappa_{b'}^{a'} = X_c^{a'} X_{b'}^d \delta_d^c = X_c^{a'} X_{b'}^c = \delta_b^a. \qquad (1.6.1)$$

Thus $\boldsymbol{\kappa}$ has the same components relative to any basis. This special tensor is called the *Kronecker tensor*, and it is customary to use δ_b^a rather than κ_b^a for its components.

A tensor $\boldsymbol{\tau}$ of type $(2, 0)$ is called *symmetric* if $\boldsymbol{\tau}(\boldsymbol{\alpha}, \boldsymbol{\beta}) = \boldsymbol{\tau}(\boldsymbol{\beta}, \boldsymbol{\alpha})$ for all $\boldsymbol{\alpha}, \boldsymbol{\beta} \in T^*$. In terms of components this is equivalent to the statement that, relative to any basis, $\tau^{ab} = \tau^{ba}$ (see Exercise 1.6.1).

One may similarly define symmetric tensors of type $(0, 2)$, and tensors of type (r, s) which are symmetric in any pair of superscripts or subscripts.

The notion of *skew-symmetry* (or *antisymmetry*) is similarly defined. If a tensor τ of type $(2, 0)$ satisfies $\tau(\alpha, \beta) = -\tau(\beta, \alpha)$ for all $\alpha, \beta \in T^*$, then it is *skew-symmetric*. In terms of components we have $\tau^{ab} = -\tau^{ba}$, relative to any basis. As with symmetry, the definition may be extended to apply to any pair of superscripts or subscripts of a type (r, s) tensor.

Exercises 1.6

1. Verify that if τ is a type $(2, 0)$ symmetric tensor as defined above, then its components relative to any basis satisfy $\tau^{ab} = \tau^{ba}$.
2. If $\sigma_{ab} = \sigma_{ba}$ and $\tau^{ab} = -\tau^{ba}$ for all a, b, prove that $\sigma_{ab}\tau^{ab} = 0$.
3. Show that any type $(2, 0)$ (or type $(0, 2)$) tensor may be expressed as the sum of a symmetric and a skew-symmetric tensor.

1.7   Metric tensors

So far we have not discussed metric properties such as the length of a vector, or the angle between vectors. In order to do so we need the concept of a metric tensor, which we now define.

A *metric tensor* $\mathbf{g}$ is a symmetric type $(0, 2)$ tensor which is non-singular in the sense that $\mathbf{g}(\lambda, \alpha) = 0$ for all $\alpha \in T$ implies that $\lambda = \mathbf{0}$. The symmetry condition implies that its components g_{ab} satisfy $g_{ab} = g_{ba}$, while the non-singularity condition implies that the matrix $[g_{ab}]$ is non-singular, (i.e. $[g_{ab}]$ has an inverse, or equivalently, $|g_{ab}| \neq 0$. See Exercise 1.7.1).

In the language of vector-space theory, a metric tensor provides T with an *inner product* $\langle \lambda, \mu \rangle$ of vectors λ, $\mu \in T$ defined by

$$\langle \lambda, \mu \rangle \equiv \mathbf{g}(\lambda, \mu) = g_{ab}\lambda^a\mu^b, \tag{1.7.1}$$

for all $\lambda, \mu \in T$, but in that context it is more usual to replace the non-singularity condition by one of positive definiteness, namely that for any $\lambda \in T$, $\mathbf{g}(\lambda, \lambda) \geq 0$, with $\mathbf{g}(\lambda, \lambda) = 0$ if and only if $\lambda = \mathbf{0}$. However, the needs of relativity require us to adopt the weaker condition of non-singularity and so admit indefinite metric tensors, which, as we shall see, yields metric properties which appear somewhat bizarre. Before discussing these we shall define the contravariant metric tensor $\hat{\mathbf{g}}$ which provides T^* with an inner product.

Since the matrix $[g_{ab}]$ is non-singular, its inverse exists. Let g^{ab} be the entry in the ath row and bth column of this inverse. Then

$$g^{ab}g_{bc} = \delta^a_c, \qquad (1.7.2)$$

and because $g_{ab} = g_{ba}$ we also have that

$$g^{ab} = g^{ba}.$$

We now use the quotient theorem to show that the g^{ab} are components of a type $(2, 0)$ tensor $\hat{\mathbf{g}}$. To this end, let α^a, β^a be the components of arbitrary contravariant vectors. Then, because of the non-singularity of $[g_{ab}]$, $\lambda_a \equiv g_{ab}\alpha^b$ and $\mu_a \equiv g_{ab}\beta^b$ are the components of *arbitrary* covariant vectors. Thus for arbitrary covariant vectors $\boldsymbol{\lambda}$ and $\boldsymbol{\mu}$,

$$\begin{aligned} g^{ab}\lambda_a\mu_b &= g^{ab}g_{ac}\alpha^c g_{bd}\beta^d \\ &= \delta^a_d g_{ac}\alpha^c\beta^d \qquad \text{(on using (1.7.2))} \\ &= g_{dc}\alpha^c\beta^d, \end{aligned}$$

which is a type $(0, 0)$ tensor. So the g^{ab} with their superscripts killed off by contraction with arbitrary covariant vectors display tensor character, and the quotient theorem implies that they are the components of a type $(2, 0)$ tensor $\hat{\mathbf{g}}$. This is the *contravariant metric tensor* determined by $\mathbf{g}$, and it provides T^* with an inner product:

$$\langle \boldsymbol{\lambda}, \boldsymbol{\mu} \rangle \equiv g^{ab}\lambda_a\mu_b, \qquad \text{for } \boldsymbol{\lambda}, \boldsymbol{\mu} \in T^*.$$

A metric tensor $\mathbf{g}$ provides us with a linear transformation from T^r_s to T^{r-1}_{s+1}. For example, if $\boldsymbol{\tau} \in T^2_1$ has components $\tau^{ab}{}_c$, then $\tau_a{}^b{}_c$ defined by $\tau_a{}^b{}_c = g_{ad}\tau^{db}{}_c$ are the components of a tensor in T^1_2. The contravariant metric tensor $\hat{\mathbf{g}}$ then gives the inverse transformation from T^{r-1}_{s+1} back to T^r_s. For, continuing the example above,

$$g^{ad}\tau_d{}^b{}_c = g^{ad}g_{de}\tau^{eb}{}_c = \delta^a_e\tau^{eb}{}_c = \tau^{ab}{}_c.$$

The components $\tau_a{}^b{}_c \equiv g_{ad}\tau^{db}{}_c$ are said to be obtained from the components $\tau^{ab}{}_c$ by *lowering a suffix*, and the components $\tau^{ab}{}_c \equiv g^{ad}\tau_d{}^b{}_c$ are said to be obtained from the components $\tau_a{}^b{}_c$ by *raising a suffix*. Tensors which may be obtained from each other by raising or lowering suffixes are said to be *associated*, and it is conventional to use the same kernel letter for components, as in the example above. However, this usage is ambiguous if we have more than one metric tensor defined on T; but since this is rarely the case, opportunity for ambiguity seldom arises. (See, however, Section 5.1.) Another source of ambiguity is the fact that more than one tensor of the same type may be associated with a given tensor. For example, lowering the first superscript of the components τ^{ab} of a type $(2, 0)$ tensor yields a type $(1, 1)$ tensor which is in general different from that obtained by lowering the second superscript. The distinction between the two may be made clear by careful spacing of the suffixes:

$$\tau_a{}^b = g_{ac}\tau^{cb}, \qquad \tau^a{}_b = g_{bc}\tau^{ac}.$$

In the case of symmetric tensors this distinction is not necessary (see, for example, the Ricci tensor in Chapter 3).

Since

$$\delta_b^a = g^{ac}g_{cb} \qquad \text{and} \qquad g^{ab} = g^{ad}\delta_d^b = g^{ad}g^{bc}g_{cd},$$

the metric tensor, the contravariant metric tensor and the Kronecker tensor are associated. However, the convention of using the same kernel letter for components of associated tensors is relaxed in the case of the Kronecker tensor because of the special form its components take, and we use δ_b^a rather than g_b^a.

We are now in a position to define some metric concepts.

The *length* of a contravariant vector $\boldsymbol{\lambda}$ is denoted by $|\boldsymbol{\lambda}|$ and is defined by

$$|\boldsymbol{\lambda}| \equiv |\mathbf{g}(\boldsymbol{\lambda}, \boldsymbol{\lambda})|^{1/2} = |g_{ab}\lambda^a\lambda^b|^{1/2} = |\lambda_a\lambda^a|^{1/2}. \tag{1.7.3}$$

The modulus signs are needed because $\mathbf{g}$ may be indefinite. In this case we can have $|\boldsymbol{\lambda}| = 0$ with $\boldsymbol{\lambda} \neq \mathbf{0}$. Such vectors are called *null*. For a covariant vector $\boldsymbol{\mu}$, its length is similarly defined in terms of $\hat{\mathbf{g}}$, or equivalently as the length of the associated contravariant vector:

$$|\boldsymbol{\mu}| \equiv |g^{ab}\mu_a\mu_b|^{1/2} = |\mu^b\mu_b|^{1/2} = |g_{ab}\mu^a\mu^b|^{1/2}. \tag{1.7.4}$$

A *unit vector* is one whose length is one.

The inner product may be used to define the *angle* θ between two non-null contravariant vectors $\boldsymbol{\lambda}$, $\boldsymbol{\mu}$ via

$$\begin{aligned} \cos\theta &\equiv \mathbf{g}(\boldsymbol{\lambda}, \boldsymbol{\mu})/(|\boldsymbol{\lambda}||\boldsymbol{\mu}|) \\ &= (g_{ab}\lambda^a\mu^b)/(|g_{cd}\lambda^c\lambda^d|^{1/2}|g_{ef}\mu^e\mu^f|^{1/2}), \end{aligned} \tag{1.7.5}$$

with obvious modifications for covariant vectors. Note that in the case of an indefinite metric tensor we may have $|\cos \theta| > 1$ giving, as it were, a complex angle between the vectors.

Two vectors are *orthogonal* (or *perpendicular*) if their inner product is zero. This definition makes sense even if one or both of the vectors are null. In fact, a null vector is a non-zero vector which is orthogonal to itself. An example in relativity is the wave 4-vector (see equation (A.6.6)).

Exercises 1.7

1. Show that if $\mathbf{g} \in T_2$ is non-singular, in the sense that $\mathbf{g}(\boldsymbol{\lambda}, \boldsymbol{\alpha}) = 0$ for all $\boldsymbol{\alpha} \in T$ implies that $\boldsymbol{\lambda} = \mathbf{0}$, then $[g_{ab}]$ is a non-singular matrix.
2. Verify that $\hat{\mathbf{g}}$ is symmetric.
3. If $\mathbf{g}$ is positive definite, prove that $\cos \theta$, as defined by equation (1.7.5), satisfies $|\cos \theta| \leqslant 1$.

1.8 An example from physics: the inertia tensor

In classical mechanics, the angular momentum of a particle about the origin O is

$$\mathbf{l} \equiv \mathbf{r} \times m\mathbf{v}, \tag{1.8.1}$$

where $\mathbf{r}$ is its position vector relative to O, m its mass and $\mathbf{v}$ its velocity. If this particle is a constituent particle of a rigid body having one of its points fixed at O, then its velocity is given by $\mathbf{v} = \boldsymbol{\omega} \times \mathbf{r}$, where $\boldsymbol{\omega}$ is the angular velocity of the body. Substitution for $\mathbf{v}$ in equation (1.8.1) gives

$$\mathbf{l} = m\mathbf{r} \times (\boldsymbol{\omega} \times \mathbf{r}) = m[r^2 \boldsymbol{\omega} - (\boldsymbol{\omega} \cdot \mathbf{r})\mathbf{r}], \tag{1.8.2}$$

where $r \equiv |\mathbf{r}|$.

Now the familiar three-dimensional space of classical mechanics is a vector space, provided our vectors are localised at the origin O. In equation (1.8.2), $\mathbf{r}$ and $\mathbf{l}$ are localised at O, while $\boldsymbol{\omega}$, being a free vector, may be regarded as localised at O. Hence equation (1.8.2) is a vector equation of the space of vectors localised at O. Moreover, the space has a metric tensor defined via the usual inner product $\mathbf{a} \cdot \mathbf{b}$ of vectors $\mathbf{a}$ and $\mathbf{b}$, and this appears in equation (1.8.2).

Let us introduce an orthonormal basis (i.e. a basis of mutually orthogonal unit vectors), but label the basis vectors $\mathbf{e}_1$, $\mathbf{e}_2$, $\mathbf{e}_3$, in accordance with the notation of earlier sections, rather than the usual $\mathbf{i}$, $\mathbf{j}$, $\mathbf{k}$. Let us also use letters i, j, k, etc., from the middle of the alphabet for the range 1, 2, 3, so our basis is $\{\mathbf{e}_i\}$. From equation (1.7.1), $g_{ij} = \langle \mathbf{e}_i, \mathbf{e}_j \rangle = \mathbf{e}_i \cdot \mathbf{e}_j$, so we see that for our orthonormal basis $g_{ij} = \delta_{ij}$ (δ_{ij} being defined in a similar way to δ^i_j), but let us stick to g_{ij} for the metric tensor components. Putting $\mathbf{r} \equiv x^i \mathbf{e}_i$, $\boldsymbol{\omega} \equiv \omega^i \mathbf{e}_i$ and $\mathbf{l} \equiv l^i \mathbf{e}_i$, the component version of equation (1.8.2) is

$$I^i = m[r^2 \omega^i - (g_{jk}\omega^j x^k)x^i],$$

where $r^2 \equiv g_{jk}x^j x^k$. This may be written as

$$l^i = m(r^2 \delta^i_j - x^i x_j)\omega^j, \tag{1.8.3}$$

where $x_j \equiv g_{jk}x^k$.

If we now sum over all particles making up the rigid body, we obtain the components of the total angular momentum of the body about O as

$$L^i = I^i_j \omega^j, \tag{1.8.4}$$

where

$$I^i_j \equiv \sum_{\substack{\text{all} \\ \text{particles}}} m(r^2 \delta^i_j - x^i x_j).$$

It may be seen that at any instant I^i_j depends only on the masses of the constituent particles and their positions, and not on $\boldsymbol{\omega}$. Since $\mathbf{I}$ is a vector, the quotient theorem entitles us to conclude that the I^i_j are the components of a type $(1, 1)$ tensor $\mathbf{I}$, namely the *inertia tensor* of the body about O. Equation (1.8.4) then states that $\mathbf{L}$ is the contraction of $\mathbf{I}$ with $\boldsymbol{\omega}$.

To illustrate the tensor transformation formula (1.4.2), let us consider a body consisting of a single particle of mass m, whose position vector at the instant in question is $\mathbf{r} = \mathbf{e}_1$. To fit in with the rigid-body picture we must imagine this particle as tethered to O by a light rigid rod of unit length. Then by definition,

$$[I^i_j] = \begin{bmatrix} 0 & 0 & 0 \\ 0 & m & 0 \\ 0 & 0 & m \end{bmatrix}.$$

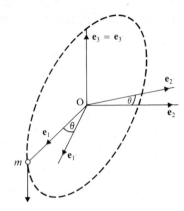

Fig. 1.1   Geometry of the basis vectors and rod/particle at the
instant in question.

If at the instant in question the rod/particle has angular velocity
$\boldsymbol{\omega} = 15\mathbf{e}_2$ (rad s^{-1}), then equation (1.8.4) gives

$$L^1 = 0, \qquad L^2 = 15m, \qquad L^3 = 0.$$

Now let us choose a new (orthonormal) basis $\{\mathbf{e}_{i'}\}$, which is
obtained from the old by a rotation through an angle θ about $\mathbf{e}_3$ (see
Fig. 1.1). From an inspection of the geometry,

$$\mathbf{e}_{1'} = \cos \theta \, \mathbf{e}_1 + \sin \theta \, \mathbf{e}_2,$$

$$\mathbf{e}_{2'} = -\sin \theta \, \mathbf{e}_1 + \cos \theta \, \mathbf{e}_2,$$

$$\mathbf{e}_{3'} = \mathbf{e}_3,$$

so the transformation matrix is

$$[X^i_{j'}] = \begin{bmatrix} \cos \theta & -\sin \theta & 0 \\ \sin \theta & \cos \theta & 0 \\ 0 & 0 & 1 \end{bmatrix}, \qquad (1.8.5)$$

with inverse

$$[X^{i'}_j] = \begin{bmatrix} \cos \theta & \sin \theta & 0 \\ -\sin \theta & \cos \theta & 0 \\ 0 & 0 & 1 \end{bmatrix}. \qquad (1.8.6)$$

Relative to the new basis we have

$$I^{i'}_{j'} = X^{i'}_k X^l_{j'} I^k_l, \qquad \omega^{i'} = X^{i'}_j \omega^j, \qquad L^{i'} = X^{i'}_j L^j,$$

giving

$$[I^{i'}_{j'}] = \begin{bmatrix} m\sin^2\theta & m\sin\theta\cos\theta & 0 \\ m\sin\theta\cos\theta & m\cos^2\theta & 0 \\ 0 & 0 & m \end{bmatrix},$$

$$[\omega^{i'}] = \begin{bmatrix} 15\sin\theta \\ 15\cos\theta \\ 0 \end{bmatrix} \quad \text{and} \quad [L^{i'}] = \begin{bmatrix} 15\,m\sin\theta \\ 15\,m\cos\theta \\ 0 \end{bmatrix}.$$

We note that

$$L^{i'} = I^{i'}_{j'}\omega^{j'},$$

as indeed it should be.

The above example indicates the validity of the tensor relationship $L^i = I^i_j\omega^j$ regardless of the choice of basis, and hence of the choice of rectilinear coordinate system with origin O, since a basis defines such a coordinate system. That is, a tensor formulation is indicative of coordinate independence. In general relativity we wish to describe the laws of physics in a coordinate-independent way, but the situation is complicated because the basic arena, spacetime, is not a vector space. As we shall see in the next chapter, spacetime is modelled by a manifold, each point of which has its own vector space, namely the tangent space at that point. These tangent spaces, together with their duals and tensor products of these, provide homes for the vectors and tensors of the theory, and it is here that they reside, rather than in spacetime itself. We shall also see how in a natural way any coordinate system provides each tangent space with a basis, and how a change of coordinates induces a change of basis of each tangent space.

Exercises 1.8

1. Verify that $L^{i'} = I^{i'}_{j'}\omega^{j'}$, as asserted above.

Problems 1

1. Suppose we set up a one-to-one correspondence between a vector space T and its dual T^*, by taking a basis $\{\mathbf{e}_a\}$ of T and saying that $\boldsymbol{\mu} \in T^*$ corresponds to $\boldsymbol{\lambda} \in T$ if the components of $\boldsymbol{\mu}$ relative to the dual basis $\{\mathbf{e}^a\}$ are the same as those of $\boldsymbol{\lambda}$

relative to $\{\mathbf{e}_a\}$. (i.e. $\boldsymbol{\lambda} \leftrightarrow \boldsymbol{\mu}$ if $\lambda^a = \mu_a$ for all a.) Show that this correspondence is not basis-independent.

Is it possible to set up a natural isomorphism between T and T^*?

2. Show that any tensor in $T \otimes U$ may be written as a sum of decomposable tensors.

3. If $\boldsymbol{\tau}$ is a symmetric type $(0, 2)$ tensor and $\boldsymbol{\lambda}$ a covariant vector, and their components satisfy

$$\tau_{bc}\lambda_a + \tau_{ca}\lambda_b + \tau_{ab}\lambda_c = 0$$

for all a, b, c, deduce that either $\boldsymbol{\tau} = \mathbf{0}$ or $\boldsymbol{\lambda} = \mathbf{0}$.

(Hint: If $\boldsymbol{\lambda} \neq \mathbf{0}$, it may be taken as the first vector of a basis.)

4. With each basis of an N-dimensional vector space there are associated N^2 numbers τ_{ab} satisfying $\tau_{ab} = \tau_{ba}$, and it is known that if λ^a are the components of an arbitrary contravariant vector $\boldsymbol{\lambda}$, then the expression $\tau_{ab}\lambda^a\lambda^b$ is invariant under a change of basis. Show that the τ_{ab} are the components of a type $(0, 2)$ tensor.

5. The type $(0, 4)$ tensor $\boldsymbol{\tau}$ satisfies $\boldsymbol{\tau}(\boldsymbol{\lambda}, \boldsymbol{\mu}, \boldsymbol{\lambda}, \boldsymbol{\mu}) = 0$ for all contravariant vectors $\boldsymbol{\lambda}$ and $\boldsymbol{\mu}$. Show that its components satisfy

$$\tau_{abcd} + \tau_{cbad} + \tau_{adcb} + \tau_{cdab} = 0.$$

6. Show, by making use of matrix theory, that if the metric tensor $\mathbf{g}$ is positive definite, it is possible to introduce a basis in which $g_{ab} = \delta_{ab}$.

If relative to two different bases $\{\mathbf{e}_a\}$ and $\{\mathbf{e}_{a'}\}$, $g_{ab} = g_{a'b'} = \delta_{ab}$, what can you say about the transformation matrix $[X^a_b]$?

Notes

1. Halmos, 1974, Ch. I.
2. Halmos, 1974, Ch. I, §8.
3. Schouten, 1954, p. 3, in particular footnote[1].

The spacetime of general relativity and paths of particles

2.0 Introduction

Einstein's general theory of relativity postulates that gravitational effects may be explained by the curvature of spacetime, and that gravity should not be regarded as a force in the conventional sense. To get a preliminary idea of what is involved it is helpful to adopt and extend an allegory of Misner, Thorne and Wheeler [1], which makes use of ants crawling over a curved surface, namely the skin of an apple.

So, with due acknowledgement to the originators of the idea, let us consider an ant crawling over the skin of an apple. The straightest path it could take would be obtained by its making its left-hand paces equal to its right-hand ones. This would clearly generate a straight-line path if it were crawling on a plane, so it is natural to adopt a path generated on a curved surface in this way as the analogue of a straight line. These paths are called *geodesics*. If the ant had inky feet, so that it left footprints, then making cuts along the left-hand and right-hand tracks would yield a thin strip of peel which could be removed. If this thin strip were laid flat on a plane it would be straight, confirming that a geodesic, as we have defined it, is the analogue of a straight line.

Suppose now that we have several ants crawling over the apple (without colliding) and each follows a geodesic path, leaving a record of its progress on the apple's skin. (A single track rather than a double one: ink on the tip of its abdomen, rather than inky feet.) If we concentrate on a portion of the apple's skin which is *so small that it may be considered flat*, then the tracks of the ants would appear as straight lines on this "flat" portion (see Fig. 2.1). If, however, we take a larger view of things, then the picture is different. For example, suppose two ants leave from nearby points on a starting line at the same time, and move with the same constant speed, following geodesics which are initially perpendicular to the starting line. Their paths would initially be

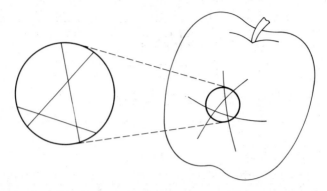

Fig. 2.1   Small portions of an apple's skin may be regarded as flat
(after Misner, Thorne and Wheeler).

parallel, but because of the curvature of the apple's skin, they
would start to converge. That is, their separation *d* does not
remain constant (see Fig. 2.2). More generally, we can see that
the relative acceleration of ants which follow neighbouring
geodesics with constant (but not necessarily equal) speeds is non-
zero, if the surface over which they are crawling is curved. In this
way curvature may be detected implicitly by what is called
geodesic deviation.

An apple is not a perfect sphere: there is a dimple caused by
the stalk. Should an ant pass near the stalk its geodesic path
would suffer a marked deflection, like that of a comet passing
near the Sun, and it would look as if the stalk attracted the ant.
However, this is not the correct interpretation. The stalk modifies
the curvature of the apple's skin in its vicinity, and this produces
geodesics which give the effect of an attraction by the stalk.

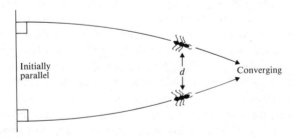

Fig. 2.2   Converging geodesics on an apple's skin.

This allegory may be interpreted in the following way. The curved surface which is the apple's skin represents the curved spacetime of Einstein's general theory, which bears the same relation to the flat spacetime of the special theory as does the apple's skin to a plane. Free particles (i.e. those moving under gravity alone, gravity no longer being a force) follow the straightest paths or geodesics in the curved spacetime, just as the ants follow geodesics on the apple's skin. Locally the spacetime of the general theory is like that of the special theory [2], but on a larger view it is curved, and this curvature may be detected implicitly by means of geodesic deviation, just as the curvature of the apple's skin may be detected by noting the convergence of neighbouring geodesics. The way in which the dimple around the stalk gives the impression of attraction corresponds to the fact that massive bodies modify the curvature of spacetime in their vicinity, and this modification affects the geodesics in such a way as to give the impression that free particles are acted on by a force, whereas in actual fact they are following the straightest paths in the curved spacetime.

The allegory may also be used to illustrate some features of coordinate systems. We remarked in the Introduction that in a local inertial frame (a freely falling, non-rotating reference system occupying a small region of spacetime) the laws of physics are those of special relativity, and in particular free particles (those moving under gravity alone) follow straight-line paths. The corresponding situation on the apple's skin is that in a region small enough to be considered flat, we may introduce a reference system whose coordinate mesh is rectilinear, with the geodesic cutting lines of the mesh at equal angles. Let us temporarily describe local coordinate systems of this kind as "good", and others as "bad" (see Fig. 2.3). If we thought the bad local coordinate system were a good one, then we would interpret Fig. 2.3(b) as in Fig. 2.4, and the geodesics would appear curved. The moral of this observation is that in a frame which is not locally inertial (a bad frame) a free particle appears to be acted on by a force, which it is conventional to call gravity, but in a local inertial frame (a good frame) gravity can be eliminated (at least locally).

Note that on a curved surface, such as the apple's skin, the good coordinate systems are valid only over a small region, and it is only if the surface is actually flat that they can be extended to

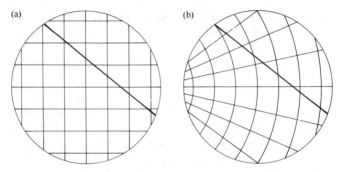

Fig. 2.3 (a) A "good" local coordinate system with geodesic; (b) a "bad" local coordinate system with geodesic.

global coordinate systems. Similarly, only if spacetime is flat (as in the special theory) can we have global inertial frames, but then there is no curvature and no gravity. It may then seem that in order to do relativity we have to restrict ourselves to limited regions of spacetime which may be covered by local inertial systems, but this is not the case. Instead we use extensive bad coordinate systems, and learn how to define physically significant concepts correctly in terms of these.

Before we can do this, we must produce the basic mathematical model for a curved spacetime, which is a four-dimensional Riemannian manifold. The necessary mathematics for this is covered in Sections 2.1–2.5, along with other related concepts such as tensor fields and geodesics. The remainder of the chapter is devoted to a closer look at geodesic motion, and a comparison

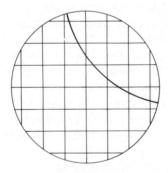

Fig. 2.4 Effect on geodesic if "bad" coordinates are mistaken for "good" ones.

with Newton's laws. It is not until Chapter 3 that we give an answer to the question of what it is that produces the curvature of spacetime.

Exercises 2.0

1. Ants follow geodesics on a surface which is an infinite cylinder. Do the geodesics deviate?

   Is the cylinder really curved?

   By considering the paths of itself and its neighbours can an ant decide whether it is on a cylinder or a plane?

2.1 Manifolds

A differentiable manifold (or manifold for short) is a generalisation of a surface in the sense that

(a) it has a dimension, N say, so that points in it may be labelled by N real coordinates $x^1, x^2, \ldots, x^N$;

(b) it can support a differentiable structure; i.e. the functions involved in changes of coordinates are differentiable.

There is, however, one important respect in which it differs from a surface, namely that it is a thing in itself, and we do not consider it embedded in some higher-dimensional euclidean space. Our formal definition is arrived at by starting with a set M, and then giving it sufficient structure so that it becomes a manifold.

Let M be a set of points and ψ a one-to-one map from a subset U of M onto an open set in $\mathbb{R}^N$. ($\mathbb{R}^N$ is the set of N-tuples $(x^1, \ldots, x^N)$, where each x^a is real. An *open* set S of $\mathbb{R}^N$ is one with the property that each point of it may be surrounded by a ball B centred on the point in question, such that B lies entirely in S. The map ψ maps U *onto* S if for each $s \in S$ we have $\psi(u) = s$ for some $u \in U$. See, for example, the text by Apostol [3] for details.) U is a *coordinate neighbourhood*, ψ is a *coordinate function*, and the pair (U, ψ) together is a *chart*. The purpose of ψ is to attach coordinates to points of U, and if P is a point in U we shall call the chart (U, ψ) a *coordinate system* about P.

Now let $\{(U_\alpha, \psi_\alpha)\}$ be a collection of charts (α being a label which distinguishes different members of the collection) with the following properties:

(a) The collection $\{U_\alpha\}$ covers M, i.e. each point of M is a member of at least one U_α.

(b) ψ_α maps U_α into $\mathbb{R}^N$ with the same N for all α.

(c) For all α, β, $\psi_\alpha \circ \psi_\beta^{-1}$ and $\psi_\beta \circ \psi_\alpha^{-1}$ are differentiable functions from $\mathbb{R}^N \to \mathbb{R}^N$ wherever they are defined. (They are defined only if U_α and U_β intersect. The inverse maps ψ_α^{-1} and ψ_β^{-1} are defined because ψ_α and ψ_β are one-to-one.)

(d) This collection is maximal in the sense that, if (U, ψ) is a chart, ψ mapping U onto an open set in $\mathbb{R}^N$ with $\psi \circ \psi_\alpha^{-1}$ and $\psi_\alpha \circ \psi^{-1}$ differentiable for all α for which they exist, then (U, ψ) belongs to the collection $\{(U_\alpha, \psi_\alpha)\}$.

A collection of charts having these properties is called an *atlas*, and M together with its atlas is an *N-dimensional differentiable manifold*.

Note that we do not claim that a manifold can be covered by a single coordinate neighbourhood (although some can), and that is why we have a whole collection of charts. Property (c) tells us how to relate things in the overlap region of two coordinate neighbourhoods.

Figure 2.5 illustrates the situation which arises when two coordinate neighbourhoods U_α and U_β intersect. The intersection is shaded, as are its images in $\mathbb{R}^N$ under ψ_α and ψ_β. The functions $\psi_\alpha \circ \psi_\beta^{-1}$ and $\psi_\beta \circ \psi_\alpha^{-1}$ are one-to-one differentiable functions which map one shaded region of $\mathbb{R}^N$ onto the other, as shown.

We should say something about the meaning of the word *differentiable*. Consider a function $f : S \to \mathbb{R}^L$, where S is an open set in $\mathbb{R}^K$. Then f is given by L component functions $f^1, \ldots, f^L$, each of which is a function of K variables, and f is said to be *differentiable of class C^r* if each f^a possesses continuous partial derivatives up to and including those of order r (r a positive integer). By a *differentiable function* we shall mean one of class C^r, where r is sufficiently large to ensure that operations which depend on the continuity of partial derivatives, such as interchanging the order of partial differentiation, are valid, and that certain concepts are well defined.

Let us consider some consequences of our definition. If (U, ψ) and (U', ψ') are charts with intersecting coordinate neighbourhoods ψ assigns coordinates $(x^1, x^2, \ldots, x^N)$, say, to points

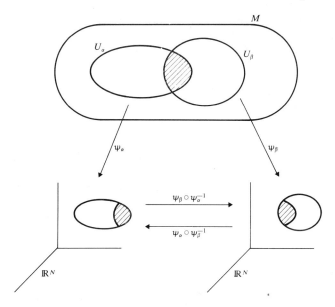

Fig. 2.5 Overlapping coordinate neighbourhoods.

in $U \cap U'$, while ψ' assigns coordinates $(x^{1'}, x^{2'}, \ldots, x^{N'})$, and the primed coordinates are given in terms of the unprimed coordinates by equations

$$x^{a'} = f^a(x^1, \ldots, x^N), \tag{2.1.1}$$

where $(f^1, \ldots, f^N) \equiv f = \psi' \circ \psi^{-1}$. $(A \cap B$ denotes the set of points common to the sets A and B.) The unprimed coordinates are similarly given in terms of the primed coordinates by equations

$$x^a = g^a(x^{1'}, \ldots, x^{N'}), \tag{2.1.2}$$

where $(g^1, \ldots, g^N) \equiv g = \psi \circ (\psi')^{-1}$. The function f and its inverse g are both one-to-one and differentiable, and it follows that the Jacobians $|\partial x^{a'}/\partial x^b|$ and $|\partial x^a/\partial x^{b'}|$ are non-zero.

Conversely, if we have a chart (U, ψ) and a system of equations of the form (2.1.1) with Jacobian $|\partial x^{a'}/\partial x^b|$ non-zero for values of x^a which are the coordinates of some point P in U, then it is possible to construct a coordinate system (U', ψ') about P whose coordinates are related to those of (U, ψ) by equations (2.1.1). (See Exercise 2.1.1. The inverse-function theorem is needed.)

We conclude this section by considering the relationship between two charts on a sphere.

Example.   Consider the unit sphere whose equation in familiar three-dimensional space is $x^2 + y^2 + z^2 = 1$. The usual polar angles θ, ϕ may be used to give a system of coordinates on it ($\theta = $ colatitude, $\phi = $ longitude). To be more precise, and to fit in with the notation introduced above, let us put $\theta \equiv x^1$, $\phi \equiv x^2$, and restrict x^1 and x^2 to satisfy $0 < x^1 < \pi$, $0 < x^2 < \pi$. In this way we produce a chart (U, ψ), where U is the hemisphere given by $y > 0$, and ψ maps U onto the open rectangle V in $\mathbb{R}^2$ given by $0 < x^1 < \pi$, $0 < x^2 < \pi$ (see Fig. 2.6).

For a second chart (U', ψ') on the sphere take U' to be the upper hemisphere given by $z > 0$, and assign to points in it the coordinate values x, y obtained by projection on to the plane $z = 0$. To fit in with the notation above let us call these coordinates $x^{1'}, x^{2'}$. The coordinate function ψ' then maps U' onto the open disc V' in $\mathbb{R}^2$ given by $(x^{1'})^2 + (x^{2'})^2 < 1$. The overlap region $U \cap U'$ is the quarter-sphere given by $y > 0$, $z > 0$, and in it the coordinates are related by

$$x^{1'} = \sin x^1 \cos x^2, \qquad x^{2'} = \sin x^1 \sin x^2$$

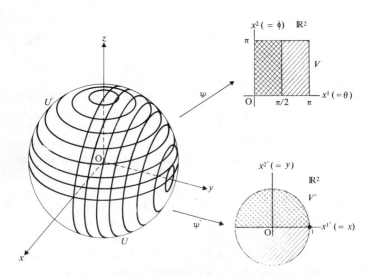

Fig. 2.6   Overlapping coordinate neighbourhoods on a sphere.

(since $x = \sin \theta \cos \phi$, $y = \sin \theta \sin \phi$). These are the equations which give the function $\psi' \circ \psi^{-1}$ mapping the doubly shaded portion of V onto the doubly shaded portion of V'. Note that

$$|\partial x^{a'}/\partial x^b| = \begin{vmatrix} \cos x^1 \cos x^2 & -\sin x^1 \sin x^2 \\ \cos x^1 \sin x^2 & \sin x^1 \cos x^2 \end{vmatrix}$$

$$= \sin x^1 \cos x^1 = \tfrac{1}{2} \sin 2x^1,$$

and in the overlap $0 < x^1 < \dfrac{\pi}{2}$, so $|\partial x^{a'}/\partial x^b| \neq 0$.

Exercises 2.1

1. In an N-dimensional manifold, (U, ψ) is a chart, and ψ maps U onto the open set V in $\mathbb{R}^N$. A differentiable function f maps a subset S of V into a set T in $\mathbb{R}^N$, and is such that the Jacobian of f is non-zero at $\psi(\mathrm{P}) \in S$. (P is a point of the manifold.) Use the inverse-function theorem [4] to show that there exists a chart (U', ψ') with $\mathrm{P} \in U'$, U' a subset of U, and $\psi' = f \circ \psi$ in U'.

2.2 The tangent space and tensor fields

The tangent space at any point of a manifold is the generalisation of the tangent plane to a surface at any point of it. With a surface embedded in three-dimensional space we can readily realise the tangent plane as a plane in the embedding space. However, a manifold has no embedding space, so we must devise some implicit means of defining the tangent space in terms of the structure available to us. The way to do this is to make use of curves in the manifold. We first define a tangent vector to a curve at a point of it, and then the tangent space at that point as the vector space of tangent vectors to curves passing through the point.

In any coordinate neighbourhood of an N-dimensional manifold M we can define a *curve* by means of N continuous functions $x^a(u)$, where u belongs to some real interval. As the *parameter* u varies we obtain the coordinates $x^a = x^a(u)$ of points on the curve. We shall confine ourselves to *regularly parametrised smooth curves*, i.e. ones which may be parametrised in such a way that at each point the derivatives $\dot{x}^a(u)$ exist, and not all of them are zero. (We shall in general use dots to denote derivatives with respect to the parameter along a curve.)

Consider now a curve γ given by $x^a(u)$, and let P be a point on it. We can choose the parametrisation so that $x^a(0) = (x^a)_P$, the coordinates of P. We define the *tangent vector* $\boldsymbol{\lambda}$ to γ at P to be the N-tuple $(\dot{x}^1(0), \dot{x}^2(0), \ldots, \dot{x}^N(0))$ [5]. Thus a curve with a given parametrisation yields a tangent vector, and conversely, each N-tuple $\boldsymbol{\lambda} \equiv (\lambda^1, \ldots, \lambda^N)$ is the tangent vector to some curve through P, e.g. that given by

$$x^a(u) = \lambda^a u + (x^a)_P. \tag{2.2.1}$$

Since we have defined a tangent vector as an N-tuple, and under obvious definitions of addition and scalar multiplication the set of N-tuples is a vector space, it follows that the set of all tangent vectors at P is a vector space. This space is the *tangent space* $T_P(M)$ of M at P. A basis of this space is $\{(\mathbf{e}_a)_P\}$, where $(\mathbf{e}_a)_P$ is the N-tuple with 1 in its ath position, and zeros elsewhere. We can then put

$$\boldsymbol{\lambda} = (\lambda^1, \ldots, \lambda^N) = \lambda^a (\mathbf{e}_a)_P,$$

and the λ^a of the N-tuple are the components of $\boldsymbol{\lambda}$ relative to the basis.

The basis vector $(\mathbf{e}_a)_P$ is in fact the tangent vector to the ath *coordinate curve* through P. That is, the curve obtained by keeping all the coordinates except the ath fixed. It is given by

$$x^b(u) = \delta_a^b u + (x^b)_P,$$

so that $x^b(0) = (x^b)_P$ as required, and $\dot{x}^b(0) = \delta_a^b$, so the N-tuple $(\dot{x}^1(0), \ldots, \dot{x}^N(0))$ has 1 in its ath position and zeros elsewhere. That is, the tangent vector to the ath coordinate curve is indeed $(\mathbf{e}_a)_P$.

The basis $\{(\mathbf{e}_a)_P\}$ whose members are tangent vectors to the coordinate curves through P is called the *natural basis* associated with the coordinate system. If we have another coordinate system about P with coordinates $x^{a'}$, then this gives rise to a new natural basis $\{(\mathbf{e}_{a'})_P\}$ of $T_P(M)$, and we can investigate the form of the transformation formula for vector components.

The curve γ through P will be given by $x^{a'}(u)$, say, in the new coordinate system, and the components of its tangent vector at P relative to the new natural basis are $\dot{x}^{a'}(0)$. But

$$\dot{x}^{a'}(0) = \left(\frac{\partial x^{a'}}{\partial x^b}\right)_P \dot{x}^b(0),$$

and since any vector $\boldsymbol{\lambda}$ in the tangent space is the tangent vector to some curve, we have in general that

$$\lambda^{a'} = (X_b^{a'})_P \lambda^b, \tag{2.2.2}$$

where $X_b^{a'} \equiv \dfrac{\partial x^{a'}}{\partial x^b}$. Comparison with equations (1.1.13) and (1.1.14) entitles us to draw the conclusion that

$$(\mathbf{e}_{a'})_P = (X_{a'}^b)_P(\mathbf{e}_b)_P,$$

where $X_{a'}^b \equiv \dfrac{\partial x^b}{\partial x^{a'}}$, for

$$\frac{\partial x^a}{\partial x^{b'}}\frac{\partial x^{b'}}{\partial x^c} = \frac{\partial x^a}{\partial x^c} = \delta_c^a,$$

showing that the matrix $[\partial x^a/\partial x^{b'}]$ is the inverse of $[\partial x^{a'}/\partial x^b]$. It is also clear that in the expression $\dfrac{\partial x^a}{\partial x^{b'}}$, the letter b should be regarded as a subscript.

To sum up, a change of coordinates about a point P of the manifold M induces a change of the natural basis of $T_P(M)$, and the fundamental matrix element involved is $X_b^{a'} \equiv \dfrac{\partial x^{a'}}{\partial x^b}$ evaluated at P.

Having defined the tangent space $T_P(M)$, we may go on to define its dual $T_P^*(M)$, and hence build up spaces $(T_s^r)_P(M)$ of type (r, s) tensors at P. We then define a type (r, s) *tensor field* on M as an assignment to each point P of M a member of $(T_s^r)_P(M)$. We denote the set of all type (r, s) tensor fields on M by $T_s^r(M)$. In particular $T^1(M) = T(M)$ is the set of contravariant vector fields on M, and $T_1(M) = T^*(M)$ is the set of covariant vector fields on M. A *scalar field* is simply a real-valued function on M. In any coordinate neighbourhood, scalar fields and components of vector and tensor fields may be regarded as functions of the coordinates. Under a change of coordinates the components of a type (r, s) tensor field therefore transform according to equation (1.4.2), where $X_b^{a'} \equiv \dfrac{\partial x^{a'}}{\partial x^b}$ and $X_{b'}^a \equiv \dfrac{\partial x^a}{\partial x^{b'}}$. It is convenient to denote the partial derivatives in this way and to extend the notation by putting $\dfrac{\partial^2 x^{a'}}{\partial x^b \partial x^c} \equiv X^{a'}_{bc}$, etc. This shorthand will be used throughout the remainder of the book.

We shall restrict our attention to *differentiable* tensor fields, i.e. those whose components are differentiable functions of the co-ordinates. In order that this concept be well defined, the differentiability class of the functions involved in the definition of M must be at least C^2. We now give a couple of examples.

As remarked above, a scalar field ϕ on M may be regarded as a function of the coordinates, so in any coordinate neighbourhood we have N real functions $\partial\phi/\partial x^a$. Then on changing coordinates, we have N new functions $\partial\phi/\partial x^{a'}$, and

$$\frac{\partial\phi}{\partial x^{a'}} = \frac{\partial\phi}{\partial x^b} \frac{\partial x^b}{\partial x^{a'}} = X^b_{a'} \frac{\partial\phi}{\partial x^b}, \qquad (2.2.3)$$

showing that $\partial\phi/\partial x^a$ are the components of a covariant vector field. This vector field is the *gradient* of the scalar field ϕ.

An example of a contravariant vector is furnished by the coordinate differences between two neighbouring points of M. If these differences δx^a are small, then to first order we have

$$\delta x^{a'} = \frac{\partial x^{a'}}{\partial x^b} \delta x^b = X^{a'}_b \delta x^b, \qquad (2.2.4)$$

showing that the small coordinate differences are the components of a contravariant vector. There is, therefore, a sense in which a small neighbourhood of a point P in M may be identified with a small region of the tangent space at P. This is analogous to ignoring the small deviation of a surface from its tangent plane near to the point of contact.

So far we have said nothing about metric properties. In order that we may do so, the manifold must carry a (differentiable) metric tensor field **g** which provides each tangent space $T_P(M)$ with a metric tensor, so enabling us to define lengths of vectors, etc., in each tangent space. In some cases a manifold inherits a metric tensor field in a natural way (for example, a surface inherits a metric structure from its embedding space), but in general it is something which has to be put on the manifold. It is always possible to do this in the positive-definite case, but for an indefinite metric tensor field to exist certain global topological requirements must be fulfilled. We shall not, however, go any further into this, since in the main we are concerned only with local results. We shall therefore assume that our manifolds are endowed with a metric tensor field **g**, so that metric properties may be discussed. A manifold which is so endowed is called a *Riemannian manifold.* If the metric tensor field is indefinite, the descriptions *pseudo-Riemannian* or *semi-Riemannian* are sometimes used, but we shall not bother to make the distinction.

A metrical definition we can make at this stage is that of the length of a curve in *M*. If the curve γ with end-points P_1 and P_2 lies entirely within a coordinate neighbourhood and is given by $x^a(u)$, $u_1 \leqslant u \leqslant u_2$, then the *length* of γ is defined to be

$$s(P_1, P_2) \equiv \int_{u_1}^{u_2} |g_{ab}\dot{x}^a\dot{x}^b|^{1/2}\, \mathrm{d}u. \tag{2.2.5}$$

It is clear that this definition is coordinate-independent, but not so clear that it does not depend on the way in which the curve is parametrised (see Exercise 2.2.2). If the metric tensor is indefinite then $g_{ab}\dot{x}^a\dot{x}^b$ may be negative, hence the need for the modulus signs. A further aspect of indefiniteness in the metric tensor is that we may have a curve whose tangent vector at each point of it is null, so that $g_{ab}\dot{x}^a\dot{x}^b = 0$ at every point, giving a curve of zero length. Such a curve is called a *null curve.*

Note that we only define lengths of curves, and make no attempt to define the distance between a pair of arbitrary points in *M*. We can, however, define the distance δs between nearby points whose coordinate differences are small. These can be regarded as points on a curve given by parameter values whose difference δu is small, and since to first order $\delta x^a = \dot{x}^a\,\delta u$, the definition yields $\delta s^2 = |g_{ab}\,\delta x^a\,\delta x^b|$. The infinitesimal version of this is

$$\mathrm{d}s^2 = |g_{ab}\,\mathrm{d}x^a\,\mathrm{d}x^b|, \tag{2.2.6}$$

(often written without the modulus signs), and $g_{ab}\,\mathrm{d}x^a\,\mathrm{d}x^b$ is often referred to as the *line element.*

In subsequent sections we shall work entirely in terms of components relative to natural bases, and we shall adopt a much-used convention, which is to *confuse a tensor with its components.* For example, we shall frequently refer to a tensor τ^{ab} rather than to a tensor $\boldsymbol{\tau}$ with components τ^{ab}.

Exercises 2.2

1. Within a coordinate neighbourhood a regularly parametrised smooth curve is given by $x^a(u)$, $a \leq u \leq b$. If the parameter is changed to $u' = f(u)$, where f is a differentiable function whose derivative is nowhere zero for $a \leq u \leq b$, show that u' is also a regular parameter, and that the tangent vector dx^a/du' is proportional to dx^a/du.

2. Show that the definition of the length of a curve given by equation (2.2.5) is independent of the (regular) parameter used.

2.3 Parallelism, differentiation and connections

Suppose that in a coordinate neighbourhood U of a manifold M we have a vector field $\lambda^a(u)$ defined along a curve γ given by $x^a(u)$ (rather than a vector field defined throughout U or M). The N quantities $\dot{\lambda}^a \equiv d\lambda^a/du$ are not the components of a vector, for if U' is another coordinate neighbourhood containing γ, then the corresponding quantities $\dot{\lambda}^{a'}$ are given by

$$\dot{\lambda}^{a'} = d(X_b^{a'}\lambda^b)/du = X_b^{a'}\dot{\lambda}^b + X_{cb}^{a'}\dot{x}^c\lambda^b, \tag{2.3.1}$$

and the term involving $X_{cb}^{a'} \equiv \partial^2 x^{a'}/\partial x^b \partial x^c$ would be absent if the $\dot{\lambda}^a$ were the components of a vector. The reason for the presence of this term is that in the defining equation

$$\dot{\lambda}^a \equiv \lim_{\delta u \to 0} \{[\lambda^a(u + \delta u) - \lambda^a(u)]/\delta u\}, \tag{2.3.2}$$

we take differences of components at different points of γ, and because in general $(X_b^{a'})_u \neq (X_b^{a'})_{u+\delta u}$ these differences in components are not the components of a vector (at either of the points in question). In the limit the difference between $(X_b^{a'})_u$ and $(X_b^{a'})_{u+\delta u}$ shows up as $X_{cb}^{a'}$. For differentiation to yield a vector, we must take component differences at the same point of γ. We can do this if we have a notion of parallelism between neighbouring tangent spaces.

Let P have coordinates x^a, and let Q be a neighbouring point with coordinates $x^a + \delta x^a$. If we have a one-to-one correspondence between $T_P(M)$ and $T_Q(M)$, we can call corresponding vectors *parallel*. It is natural to demand that such a correspondence (a) is linear, and (b) reduces to the identity when Q coincides with P. If we denote by $\lambda^a + \delta^*\lambda^a$ the vector at Q parallel to the vector λ^a at P, then (a) implies that

$$\lambda^a + \delta^*\lambda^a = Y^a_b\lambda^b, \qquad (2.3.3)$$

where the matrix elements Y^a_b depend only on P and Q, and not on λ^a, while (b) is satisfied if, to first order in δx^a,

$$Y^a_b = \delta^a_b - \Gamma^a_{bc}\delta x^c, \qquad (2.3.4)$$

where the Γ^a_{bc} depend only on P. (The reason for the minus sign is for reconciliation with traditional notation.) If, therefore, we have N^3 quantities Γ^a_{bc} defined at each point of our coordinate neighbourhood, we can say that the vector $\lambda^a + \delta^*\lambda^a$ at the point Q with coordinates $x^a + \delta x^a$ is parallel to the vector λ^a at the point P with coordinates x^a, if

$$\lambda^a + \delta^*\lambda^a = (\delta^a_b - \Gamma^a_{bc}\,\delta x^c)\lambda^b, \qquad (2.3.5)$$

i.e. if

$$\delta^*\lambda^a = -\Gamma^a_{bc}\lambda^b\delta x^c. \qquad (2.3.6)$$

The quantities Γ^a_{bc} are called *connection coefficients*. We shall see later that for a Riemannian manifold there is a natural way of defining them in terms of the metric tensor field, but for the moment we take them as given, and show how use is made of them to define operations of differentiation on tensor fields which yield tensor fields.

Let us now return to the situation of a vector field $\lambda^a(u)$ defined along a curve γ. Let P be the point with parameter value u, and Q a neighbouring point with parameter value $u + \delta u$. Then

$$\lambda^a(u + \delta u) = \lambda^a(u) + \delta\lambda^a$$

is a vector at Q, as is the vector $\lambda^a(u) + \delta^*\lambda^a$, which is parallel to $\lambda^a(u)$ at P. The difference $\delta\lambda^a - \delta^*\lambda^a$ is also a vector at Q, as is $[\delta\lambda^a - \delta^*\lambda^a]/\delta u$. We define the *absolute derivative* $D\lambda^a/du$ of $\lambda^a(u)$ along γ to be

$$\frac{D\lambda^a}{du} \equiv \lim_{\delta u \to 0} \frac{\delta\lambda^a - \delta^*\lambda^a}{\delta u}.$$

In the limit Q→P, so this is a vector at P. Since to first order, $\delta^*\lambda^a = -\Gamma^a_{bc}\lambda^b(u)\,\delta x^c$, we see that

$$\frac{D\lambda^a}{du} \equiv \frac{d\lambda^a}{du} + \Gamma^a_{bc}\lambda^b\frac{dx^c}{du}, \qquad (2.3.7)$$

where all quantities are evaluated at the same point P of γ. Thus the absolute derivative of a vector field λ^a along a curve γ (which is a vector field along γ) involves not only the total derivative $d\lambda^a/du$ (which is not a vector field along γ), but also the connection coefficients Γ^a_{bc} and the tangent vector field dx^a/du to γ.

If U' is another coordinate neighbourhood, then the connection coefficients $\Gamma^{a'}_{b'c'}$ will be different functions of the coordinates $x^{a'}$. We can obtain the transformation formula which gives $\Gamma^{a'}_{b'c'}$ in terms of Γ^a_{bc} in the overlap $U \cap U'$ (assumed non-empty) by considering absolute derivatives along curves lying in $U \cap U'$. For such a curve the absolute derivative is a vector, so

$$\frac{D\lambda^{a'}}{du} = X^{a'}_d \frac{D\lambda^d}{du}.$$

That is,

$$\dot{\lambda}^{a'} + \Gamma^{a'}_{b'c'}\lambda^{b'}\dot{x}^{c'} = X^{a'}_d(\dot{\lambda}^d + \Gamma^d_{ef}\lambda^e\dot{x}^f).$$

On putting $\lambda^a = X^a_{b'}\lambda^{b'}$ and $\dot{x}^f = X^f_{c'}\dot{x}^{c'}$ in the right-hand side, and using $X^{a'}_d X^d_{b'} = \delta^a_b$, this reduces to

$$(\Gamma^{a'}_{b'c'} - \Gamma^d_{ef}X^{a'}_d X^e_{b'} X^f_{c'} - X^{a'}_d X^d_{c'b'})\lambda^{b'}\dot{x}^{c'} = 0.$$

Since at any point P in $U \cap U'$ the above is true for arbitrary vector fields along arbitrary curves through P (and so for arbitrary $\lambda^{b'}$ and $\dot{x}^{c'}$ at P), and since the bracketed expression does not depend on the vector field nor on the curve, we may conclude that it is zero. Hence the required transformation formula is

$$\Gamma^{a'}_{b'c'} = \Gamma^d_{ef}X^{a'}_d X^e_{b'} X^f_{c'} + X^{a'}_d X^d_{c'b'}, \tag{2.3.8}$$

and we see that the connection coefficients are *not* the components of a tensor.

A *connection* (sometimes called an *affine* or *linear connection*) is defined on M if its coefficients are given (as functions of the coordinates) in each member of a collection of coordinate neighbourhoods which cover M, and if, in addition, transformation formulae of the kind (2.3.8.) are satisfied in each overlap region. Since the metric tensor field provides a natural candidate for a connection in a Riemannian manifold, we shall not investigate the general question of the existence of connections.

Having defined the absolute derivative of contravariant vector fields along a curve, we go on to define the absolute derivative of general tensor fields along curves. There are two approaches which may be taken. The first is to extend the notion of parallelism between neighbouring tangent spaces $T_P(M)$ and $T_Q(M)$ to one between $(T_s^r)_P(M)$ and $(T_s^r)_Q(M)$, while the second is to demand that the operation of absolute differentiation satisfies certain reasonable conditions which allow us to extend the concept to general tensor fields along curves. We shall take the latter course, and impose the following conditions on the differential operator D/du applied to tensor fields defined along a curve parametrised by u:

(a) For a tensor field $\boldsymbol{\tau}$, $D\boldsymbol{\tau}/du$ is a tensor field of the same type as $\boldsymbol{\tau}$.

(b) D/du is linear, i.e. for all tensor fields $\boldsymbol{\sigma}$, $\boldsymbol{\tau}$ of the same type, $D(\boldsymbol{\sigma}+\boldsymbol{\tau})/du = D\boldsymbol{\sigma}/du + D\boldsymbol{\tau}/du$, and for any tensor field $\boldsymbol{\tau}$ and any constant k, $D(k\boldsymbol{\tau})/du = k(D\boldsymbol{\tau}/du)$.

(c) For any scalar field ϕ, $D\phi/du = d\phi/du$.

(d) D/du obeys Leibniz' rule with respect to tensor products, i.e. for all tensor fields $\boldsymbol{\sigma}$, $\boldsymbol{\tau}$, $D(\boldsymbol{\sigma}\otimes\boldsymbol{\tau})/du = (D\boldsymbol{\sigma}/du)\otimes\boldsymbol{\tau} + \boldsymbol{\sigma}\otimes(D\boldsymbol{\tau}/du)$.

(e) D/du commutes with contraction.

If μ_a is a covariant vector field along a curve γ, then for any contravariant vector field λ^a along γ, $\lambda^a\mu_a$ is a scalar field, so

$$d(\lambda^a\mu_a)/du = D(\lambda^a\mu_a)/du.$$

Conditions (d) and (e) imply that Leibniz' rule holds for contracted products, so

$$\frac{d\lambda^a}{du}\mu_a + \lambda^a\frac{d\mu_a}{du} = \frac{D\lambda^a}{du}\mu_a + \lambda^a\frac{D\mu_a}{du}$$

$$= \mu_a\left(\frac{d\lambda^a}{du} + \Gamma^a_{bc}\lambda^b\frac{dx^c}{du}\right) + \lambda^a\frac{D\mu_a}{du},$$

which implies that

$$\lambda^a\frac{D\mu_a}{du} = \lambda^a\frac{d\mu_a}{du} - \Gamma^a_{bc}\lambda^b\frac{dx^c}{du}\mu_a.$$

Since this holds for arbitrary vector fields λ^a, we deduce by taking for λ^a the standard vector with components $\lambda^a = \delta^a_d$ that

$$\mathrm{D}\mu_d/\mathrm{d}u \equiv \dot{\mu}_d - \Gamma^a_{dc}\mu_a \dot{x}^c, \qquad (2.3.9)$$

and in this way our conditions yield the absolute derivative of a covariant vector field.

We know from Problem 1.2 that a type $(2, 0)$ tensor may be written as a sum of decomposable tensors, so as far as absolute differentiation of a type $(2, 0)$ tensor field is concerned, this fact together with condition (b) implies that there is no loss of generality in taking $\tau^{ab} = \lambda^a \mu^b$, where λ^a, μ^a are contravariant vector fields along the curve. Then using condition (d) we have

$$\mathrm{D}\tau^{ab}/\mathrm{d}u = \mathrm{D}(\lambda^a \mu^b)/\mathrm{d}u = (\mathrm{D}\lambda^a/\mathrm{d}u)\mu^b + \lambda^a(\mathrm{D}\mu^b/\mathrm{d}u).$$

Inserting appropriate expressions for $\mathrm{D}\lambda^a/\mathrm{d}u$ and $\mathrm{D}\mu^b/\mathrm{d}u$, and recombining $\lambda^a \mu^b$ as τ^{ab} results in

$$\mathrm{D}\tau^{ab}/\mathrm{d}u \equiv \dot{\tau}^{ab} + \Gamma^a_{cd}\tau^{cb}\dot{x}^d + \Gamma^b_{cd}\tau^{ac}\dot{x}^d, \qquad (2.3.10)$$

which gives the absolute derivative of a type $(2, 0)$ tensor field.

Similarly one may show that for a type $(0, 2)$ tensor field,

$$\mathrm{D}\tau_{ab}/\mathrm{d}u \equiv \dot{\tau}_{ab} - \Gamma^c_{ad}\tau_{cb}\dot{x}^d - \Gamma^c_{bd}\tau_{ac}\dot{x}^d, \qquad (2.3.11)$$

and for a type $(1, 1)$ tensor field,

$$\mathrm{D}\tau^a_b/\mathrm{d}u \equiv \dot{\tau}^a_b + \Gamma^a_{cd}\tau^c_b\dot{x}^d - \Gamma^c_{bd}\tau^a_c\dot{x}^d, \qquad (2.3.12)$$

(see Exercise 2.3.3).

The pattern should now be clear. The absolute derivative of a type (r, s) tensor field $\tau^{a_1 \ldots a_r}_{b_1 \ldots b_s}$ along a curve γ is given by the sum of the total derivative $\dot{\tau}^{a_1 \ldots a_r}_{b_1 \ldots b_s}$ of its components, r terms of the form $\Gamma^{a_k}_{cd}\tau^{\ldots c \ldots}_{\ldots}\dot{x}^d$ and s terms of the form $-\Gamma^c_{b_k d}\tau^{\ldots}_{\ldots c \ldots}\dot{x}^d$. For example

$$\mathrm{D}\tau^{ab}_c/\mathrm{d}u \equiv \dot{\tau}^{ab}_c + \Gamma^a_{de}\tau^{db}_c\dot{x}^e + \Gamma^b_{de}\tau^{ad}_c\dot{x}^e - \Gamma^d_{ce}\tau^{ab}_d\dot{x}^e. \qquad (2.3.13)$$

A vector field along a curve γ which is such that neighbouring vectors are parallel is called a *parallel field of vectors* along γ. This means that in the notation above $\delta\lambda^a = \delta^*\lambda^a$, which implies that the condition for λ^a to be a parallel field of vectors is

$$\mathrm{D}\lambda^a/\mathrm{d}u = 0. \qquad (2.3.14)$$

This equation may be regarded as a system of differential equations for determining a parallel field of vectors. Suppose we specify $\lambda^a(u)$ at the point with parameter u_0, $\lambda^a(u_0) \equiv \lambda_0^a$ say. Writing equation (2.3.14) as

$$\frac{d\lambda^a}{du} + \Gamma_{bc}^a \lambda^b \frac{dx^c}{du} = 0 \qquad (2.3.15)$$

shows that it is a linear system of first-order equations for the N functions $\lambda^a(u)$, and given the initial condition $\lambda^a(u_0) = \lambda_0^a$, these may be solved to give a parallel field of vectors $\lambda^a(u)$ along γ. This field is said to be obtained by the *parallel transport* of λ_0^a along γ.

Obtaining a parallel field of vectors along a curve by solving equations (2.3.15) is equivalent to integrating infinitesimal parallel displacements, and since the notion of parallelism is only defined for vectors at neighbouring points, it is not surprising that in general parallel transport is path-dependent. We shall say more about this later. By taking the analogue of equation (2.3.14) for tensor fields of any type, we can clearly extend the notions of parallel fields and parallel transport to tensors of any type. Parallel transport of tensors is in general path-dependent, but scalar fields are an exception, for $D\phi/du = 0$ implies that $d\phi/du = 0$, which implies that ϕ is constant along the curve.

For absolute differentiation, the tensor fields involved need only be defined along the curve in question. If we have a tensor field defined throughout M (or throughout a coordinate neighbourhood of M), then we can define another kind of derivative, namely the covariant derivative.

Suppose, for example, we have a contravariant vector field λ^a defined throughout a coordinate neighbourhood U. If γ is a curve in U, we can restrict λ^a to γ, and define its absolute derivative

$$D\lambda^a/du \equiv \dot{\lambda}^a + \Gamma_{bc}^a \lambda^b \dot{x}^c \qquad (2.3.16)$$

But $\dot{\lambda}^a = \dfrac{\partial \lambda^a}{\partial x^c} \dot{x}^c$, so this may be written

$$\frac{D\lambda^a}{du} = \left(\frac{\partial \lambda^a}{\partial x^c} + \Gamma_{bc}^a \lambda^b \right) \dot{x}^c.$$

The bracketed expression on the right of this last equation does not depend on γ but only on the components λ^a and their derivatives at the point in question, and the equation is true for arbitrary tangent vectors $\dot{x}^a$ at the point in question. The usual argument involving the quotient theorem entitles us to deduce that $\dfrac{\partial \lambda^a}{\partial x^c} + \Gamma^a_{bc} \lambda^b$ are the components of a type $(1, 1)$ tensor field. This tensor field is the *covariant derivative* of the vector field λ^a, and we denote it by $\lambda^a{}_{;c}$.

It is convenient at this point to introduce some more notation. We shall abbreviate $\partial/\partial x^a$ to ∂_a, and also use a comma followed by a subscript a written after the object on which it is acting to mean the same thing. So the covariant derivative of λ^a may be written as

$$\lambda^a{}_{;c} \equiv \partial_c \lambda^a + \Gamma^a_{bc} \lambda^b = \lambda^a{}_{,c} + \Gamma^a_{bc} \lambda^b . \qquad (2.3.17)$$

The notation extends naturally to repeated derivatives. For example we write $\partial^2 \lambda^a/\partial x^b \partial x^c$ as $\partial_b \partial_c \lambda^a$ or $\lambda^a{}_{,cb}$.

Returning now to covariant differentiation, we see that the argument above may be applied to a type (r, s) tensor field so as to define its covariant derivative, and it is clear that the resulting tensor field is of type $(r, s + 1)$. For example, for a covariant vector field μ_a, the covariant derivative is

$$\mu_{a;c} \equiv \mu_{a,c} - \Gamma^b_{ac} \mu_b , \qquad (2.3.18)$$

and this is a type $(0, 2)$ tensor field.

From the way in which we have defined the covariant derivative it follows that it satisfies conditions analogous to (a)–(e) stipulated for the absolute derivative. For example, the analogue of (c) is that the covariant derivative $\phi_{;a}$ of a scalar field ϕ is $\phi_{,a}$, the gradient of ϕ. Covariant differentiation bears the same relation to partial differentiation as absolute differentiation does to total differentiation. It is a modification which is necessary to produce an operation which when applied to a tensor field results in a tensor field. However, there is one important respect in which covariant differentiation differs from partial differentiation. With suitable conditions on the differentiability of the components $\tau^{a\cdots}_{b\cdots}$ of a tensor field we know that $\tau^{a\cdots}_{b\cdots,cd} = \tau^{a\cdots}_{b\cdots,dc}$, but as we shall see in Section 3.2, in general $\tau^{a\cdots}_{b\cdots;cd} \neq \tau^{a\cdots}_{b\cdots;dc}$. Here we have introduced the notation $\tau^{a\cdots}_{b\cdots;cd} \equiv (\tau^{a\cdots}_{b\cdots;c})_{;d}$ for repeated covariant derivatives.

Exercises 2.3

1. Show that an alternative form for the transformation formula (2.3.8) is

$$\Gamma^{a'}_{b'c'} = \Gamma^{d}_{ef} X^{a'}_{d} X^{e}_{b'} X^{f}_{c'} - X^{e}_{b'} X^{f}_{c'} X^{a'}_{ef}.$$

2. If Γ^{a}_{bc} and $\tilde{\Gamma}^{a}_{bc}$ are the coefficients of two different connections on a manifold, show that $\tau^{a}_{bc} \equiv \Gamma^{a}_{bc} - \tilde{\Gamma}^{a}_{bc}$ are the components of a type $(1, 2)$ tensor field.

3. Derive the results (2.3.11) and (2.3.12), using a method similar to that used in deriving the result (2.3.10).

4. Show that the absolute and covariant derivatives of the Kronecker tensor field are zero, (i.e. show that $D\delta^{a}_{b}/du = 0$ and $\delta^{a}_{b;c} = 0$).

2.4 The metric connection

If in ordinary three-dimensional euclidean space we take two vectors $\boldsymbol{\lambda}_0$, $\boldsymbol{\mu}_0$ at some point on a curve, and transport them along the curve keeping their lengths and directions constant, then we obtain two parallel fields of vectors $\boldsymbol{\lambda}$, $\boldsymbol{\mu}$ along the curve, and it is clear that the inner product $\boldsymbol{\lambda} \cdot \boldsymbol{\mu}$ (the ordinary dot product of vector algebra) is constant along the curve. In a Riemannian manifold we have an inner product defined, and it seems reasonable to require that the analogue of the above situation in euclidean space should hold. This requirement, together with the requirement that the connection be symmetric, leads to a unique connection whose coefficients are determined by the metric tensor field and its derivatives. By a *symmetric connection* we mean one whose coefficients satisfy $\Gamma^{a}_{bc} = \Gamma^{a}_{cb}$. This definition is coordinate-independent provided the differentiability class of the manifold is such that for changes of coordinates $X^{d}_{c'b'} = X^{d}_{b'c'}$, as the transformation formula (2.3.8) shows.

Suppose then that our connection is symmetric, and is such that if λ^{a} and μ^{a} are any parallel vector fields along any curve, then the inner product $g_{ab}\lambda^{a}\mu^{b}$ is constant along that curve. That is, for any curve parametrised by u, $D\lambda^{a}/du = D\mu^{a}/du = 0$ implies that $d(g_{ab}\lambda^{a}\mu^{b})/du = 0$. But

$$d(g_{ab}\lambda^{a}\mu^{b})/du = D(g_{ab}\lambda^{a}\mu^{b})/du$$

$$= (Dg_{ab}/du)\lambda^{a}\mu^{b} + g_{ab}(D\lambda^{a}/du)\mu^{b} + g_{ab}\lambda^{a}(D\mu^{b}/du),$$

and

$$D g_{ab}/du = g_{ab;c}\dot{x}^c,$$

so our condition requires that

$$g_{ab;c}\lambda^a\mu^b\dot{x}^c = 0.$$

Since at any point this must hold for all vectors λ^a, μ^a and all tangent vectors $\dot{x}^a$, our condition is equivalent to $g_{ab;c} = 0$, or

$$\partial_c g_{ab} = \Gamma^d_{ac}g_{db} + \Gamma^d_{bc}g_{ad}. \qquad (2.4.1)$$

Relabelling we have

$$\partial_a g_{bc} = \Gamma^d_{ba}g_{dc} + \Gamma^d_{ca}g_{bd}, \qquad (2.4.2)$$

and

$$\partial_b g_{ca} = \Gamma^d_{cb}g_{da} + \Gamma^d_{ab}g_{cd}. \qquad (2.4.3)$$

Taking $(2.4.1)+(2.4.2)-(2.4.3)$ and using the symmetry of the connection and the metric tensor gives

$$2\Gamma^d_{ca}g_{db} = \partial_c g_{ab} + \partial_a g_{bc} - \partial_b g_{ca}.$$

Contracting with $\frac{1}{2}g^{eb}$ then gives

$$\Gamma^e_{ca} \equiv \tfrac{1}{2}g^{eb}(\partial_c g_{ba} + \partial_a g_{cb} - \partial_b g_{ca}), \qquad (2.4.4)$$

and our condition does indeed determine the connection coefficients in terms of g_{ab} and its derivatives.

The statement that there exists a unique symmetric connection which preserves inner products under parallel transport is known as the *fundamental theorem of Riemannian geometry*. This connection is called the *metric* (or *Riemannian*) *connection*.

If we put

$$\Gamma_{abc} \equiv \tfrac{1}{2}(\partial_b g_{ac} + \partial_c g_{ba} - \partial_a g_{bc}), \qquad (2.4.5)$$

then equation (2.4.4) gives

$$\Gamma^a_{bc} = g^{ad}\Gamma_{dbc}, \qquad (2.4.6)$$

and a short calculation shows that

$$\Gamma_{abc} = g_{ad}\Gamma^d_{bc}. \qquad (2.4.7)$$

Also, equation (2.4.1) may be written

$$g_{ab,c} = \partial_c g_{ab} = \Gamma_{bac} + \Gamma_{abc}. \qquad (2.4.8)$$

The traditional names for Γ_{abc} and Γ^a_{bc} as defined by equations (2.4.5) and (2.4.6) are *Christoffel symbols of the first and second kinds* respectively, and the notation $\Gamma_{abc} \equiv [bc, a]$, $\Gamma^a_{bc} \equiv \{^a_{bc}\}$ is often used.

The essential property of the metric connection is that it gives $g_{ab;c} = 0$, and hence $Dg_{ab}/du = 0$ along curves. It is a simple matter (see Exercise 2.4.2) to show that $g^{ab}_{;c} = 0$ and $Dg^{ab}/du = 0$. We also have $\delta^a_{b;c} = 0$ and $D\delta^a_b/du = 0$, but this is true for any connection (see Exercise 2.3.4). *From here on Γ^a_{bc} will always denote the coefficients of the metric connection.*

The metric connection allows us to extend the familiar notion of the divergence of a vector field in euclidean space to vector and tensor fields on a manifold. For a contravariant vector field λ^a we define its *divergence* to be the scalar field $\lambda^a_{;a}$. This definition is reasonable, for in a cartesian coordinate system in euclidean space $g_{ab} = \delta_{ab}$, so $\partial_c g_{ab} = 0$ giving $\Gamma^a_{bc} = 0$, and $\lambda^a_{;a}$ reduces to $\lambda^a_{,a}$. The *divergence* of a covariant vector field μ_a is defined to be that of the associated contravariant vector field $\mu^a \equiv g^{ab}\mu_b$. For a type (r, s) tensor field we may define $(r + s)$ divergences,

$$\tau^{a_1 \ldots c \ldots a_r}_{b_1 \ldots b_s}{}_{;c}, \quad (\tau^{a_1 \ldots a_r}_{b_1 \ldots c \ldots b_s} g^{cd})_{;d},$$

although these will not be distinct if the tensor field possesses symmetries.

We finish this section with a calculation which gives a useful formula for the contraction Γ^a_{ab} of the connection coefficients. If we denote $|g_{ab}|$ by g, then the cofactor of g_{ab} in this determinant is gg^{ab}. (Note that g is not a scalar: changing coordinates changes the value of g at any point.) It follows that $\partial_c g = (\partial_c g_{ab})gg^{ab}$, and using equations (2.4.8) and (2.4.6) we have

$$\partial_c g = gg^{ab}(\Gamma_{bac} + \Gamma_{abc}) = g(\Gamma^a_{ac} + \Gamma^b_{bc})$$
$$= 2g\Gamma^a_{ac}.$$

Hence

$$\Gamma^a_{ab} = \tfrac{1}{2}g^{-1}\partial_b g = \tfrac{1}{2}\partial_b \ln |g|, \tag{2.4.9}$$

the modulus signs being needed as g is not necessarily positive in the indefinite case. Alternative expressions are

$$\Gamma^a_{ab} = \partial_b \ln |g|^{1/2} \quad \text{and} \quad \Gamma^a_{ab} = |g|^{-1/2} \partial_b |g|^{1/2}. \tag{2.4.10}$$

Exercises 2.4

1. Verify that the metric connection coefficients as defined in equation (2.4.4) transform according to the formula (2.3.8).
2. Show that for the metric connection $Dg^{ab}/du = 0$ and $g^{ab}{}_{;c} = 0$.
3. Let O, with coordinates x^a_O, be any point of a coordinate neighbourhood U, and let $x^{a'}$ be defined by the equation

$$x^{a'} \equiv x^a - x^a_O + \tfrac{1}{2}(\Gamma^a_{bc})_O(x^b - x^b_O)(x^c - x^c_O),$$

where $(\Gamma^a_{bc})_O$ are the coefficients of the metric (or any other symmetric) connection evaluated at O.
Show that $(X^{a'}_d)_O = \delta^a_d$, and hence deduce from Exercise 2.1.1 that the equation defines a new coordinate system about O with $x^{a'}$ as coordinates in the new coordinate neighbourhood U'.
Show further that $(X^{a'}_{de})_O = (\Gamma^a_{de})_O$, and hence deduce from Exercise 2.3.1 that $(\Gamma^{a'}_{b'c'})_O = 0$. (This exercise shows that about any point O of a manifold we may introduce a coordinate system in which the connection coefficients vanish at O. Such coordinates are called *geodesic coordinates* with origin O, but this is something of a misnomer.)

2.5   Geodesics

Geodesics are curves in a manifold analogous to straight lines in euclidean space, and we have already introduced the idea in Section 2.0. One way of characterising a straight line is as the shortest curve between two points, and this characterisation could be used in a Riemannian manifold where the length of a curve is defined. However, it presents some technical difficulties, particularly when the metric tensor field is indefinite (as in the spacetime of general relativity), since in that case we can have curves (or sections of curves) of zero length. We therefore adopt another characterisation of a straight line, namely its *straightness*, as our guide to defining geodesics.

To make the idea clear, consider the tangent vector field to a curve in euclidean space. By a suitable choice of parametrisation (using arc-length as the parameter will do) we can arrange for all the vectors in the field to have the same length, and it is clear that only if the curve is a straight line is this field of tangent vectors a parallel field of vectors along the curve. We may therefore characterise a curve in a manifold as being a *geodesic* if there exists a parametrisation of it such that the tangent vectors $\lambda^a \equiv dx^a/du$ constitute a parallel field of vectors along the curve. Such a parameter is called an *affine parameter*. Putting $\lambda^a = \lambda^a = dx^a/du$ in equation (2.3.15) results in

$$\frac{d^2x^a}{du^2} + \Gamma^a_{bc}\frac{dx^b}{du}\frac{dx^c}{du} = 0 \qquad (2.5.1)$$

as the defining equation of an affinely parametrised geodesic.

If we introduce another parameter $v = f(u)$, then equation (2.5.1) becomes

$$\frac{d^2x^a}{dv^2} + \Gamma^a_{bc}\frac{dx^b}{dv}\frac{dx^c}{dv} = h(v)\frac{dx^a}{dv}, \qquad (2.5.2)$$

where $h(v) \equiv -\dfrac{d^2v}{du^2} \bigg/ \left(\dfrac{dv}{du}\right)^2$. Equation (2.5.2) is the defining

equation of an arbitrarily parametrised geodesic, and given an equation of this form with $h(v)$ some specified function, it is a simple matter to show that we can change the parametrisation so that it reduces to the form (2.5.1). We simply solve the differential equation

$$\frac{d^2v}{du^2} + h(v)\left(\frac{dv}{du}\right)^2 = 0 \qquad (2.5.3)$$

(see Exercise 2.5.1).

From equation (2.5.2) we can see that the new parameter v is also affine if and only if $h(v) = 0$, i.e. if and only if $d^2v/du^2 = 0$, which gives $v = Au + B$, where A, B are constants. Hence any two affine parameters u, v are connected by an equation of the form $v = Au + B$. That is, u and v are affinely related, and this is the reason for describing them as affine.

For a non-null geodesic the arc-length s may be used as a parameter, and in accordance with equation (2.5.2) the tangent vector dx^a/ds satisfies

$$\frac{D}{ds}\left(\frac{dx^a}{ds}\right) = h(s)\frac{dx^a}{ds}. \qquad (2.5.4)$$

But from equation (2.2.6) we also have $g_{ab}\dfrac{dx^a}{ds}\dfrac{dx^b}{ds} = \pm 1$, and differentiation gives

$$
\begin{aligned}
0 &= \frac{\mathrm{d}}{\mathrm{d}s}\left(g_{ab}\frac{\mathrm{d}x^a}{\mathrm{d}s}\frac{\mathrm{d}x^b}{\mathrm{d}s}\right) \\
&= \frac{\mathrm{D}}{\mathrm{d}s}\left(g_{ab}\frac{\mathrm{d}x^a}{\mathrm{d}s}\frac{\mathrm{d}x^b}{\mathrm{d}s}\right) \\
&= g_{ab}\frac{\mathrm{D}}{\mathrm{d}s}\left(\frac{\mathrm{d}x^a}{\mathrm{d}s}\right)\frac{\mathrm{d}x^b}{\mathrm{d}s} + g_{ab}\frac{\mathrm{d}x^a}{\mathrm{d}s}\frac{\mathrm{D}}{\mathrm{d}s}\left(\frac{\mathrm{d}x^b}{\mathrm{d}s}\right) \qquad (\text{as } \mathrm{D}g_{ab}/\mathrm{d}s = 0) \\
&= 2h(s)g_{ab}\frac{\mathrm{d}x^a}{\mathrm{d}s}\frac{\mathrm{d}x^b}{\mathrm{d}s} \qquad (\text{from } (2.5.4)) \\
&= \pm 2h(s).
\end{aligned}
$$

So $h(s) = 0$, showing that s is an affine parameter along any non-null geodesic, and any other affine parameter u has the form $u = As + B$.

In order to obtain the parametric equations $x^a = x^a(u)$ of an affinely parametrised geodesic, we must solve the system of differential equations (2.5.1). These equations are second-order, and require $2N$ conditions to determine a unique solution. Suitable conditions are given by specifying the coordinates x_0^a of some point on the geodesic, and the components $\dot{x}_0^a$ of the tangent vector at that point. Bearing in mind the equations (2.4.4) which define the Γ_{bc}^a, it would seem to be a complicated procedure just to set up the geodesic equations, let alone solve them. Fortunately the equations may be generated by a very neat procedure which also produces the Γ_{bc}^a as a spin-off.

Consider the *Lagrangian* $L(\dot{x}^c, x^c) \equiv \tfrac{1}{2}g_{ab}(x^c)\dot{x}^a\dot{x}^b$, which we regard as a function of $2N$ independent variables x^c and $\dot{x}^c$. The *Euler–Lagrange equations* for a Lagrangian are the equations

$$
\frac{\mathrm{d}}{\mathrm{d}u}\left(\frac{\partial L}{\partial \dot{x}^c}\right) - \frac{\partial L}{\partial x^c} = 0, \tag{2.5.5}
$$

and for the given Lagrangian these reduce to the geodesic equations (in covariant rather than contravariant form), as we now show.

Differentiating the Lagrangian we have

$$
\frac{\partial L}{\partial \dot{x}^c} = \tfrac{1}{2}g_{ab}\,\delta_c^a\dot{x}^b + \tfrac{1}{2}g_{ab}\dot{x}^a\,\delta_c^b = g_{cb}\dot{x}^b,
$$

and

$$\frac{\partial L}{\partial x^c} = \tfrac{1}{2}\partial_c g_{ab}\dot{x}^a\dot{x}^b,$$

so equations (2.5.5) are

$$\mathrm{d}(g_{cb}\dot{x}^b)/\mathrm{d}u - \tfrac{1}{2}\partial_c g_{ab}\dot{x}^a\dot{x}^b = 0,$$

or

$$g_{cb}\ddot{x}^b + \partial_a g_{cb}\dot{x}^a\dot{x}^b - \tfrac{1}{2}\partial_c g_{ab}\dot{x}^a\dot{x}^b = 0.$$

But

$$\partial_a g_{cb}\dot{x}^a\dot{x}^b = \tfrac{1}{2}\partial_a g_{cb}\dot{x}^a\dot{x}^b + \tfrac{1}{2}\partial_b g_{ca}\dot{x}^a\dot{x}^b,$$

so we have

$$g_{cb}\ddot{x}^b + \tfrac{1}{2}(\partial_a g_{cb} + \partial_b g_{ca} - \partial_c g_{ab})\dot{x}^a\dot{x}^b = 0.$$

That is, the Euler–Lagrange equations reduce to

$$g_{cb}\ddot{x}^b + \Gamma_{cab}\dot{x}^a\dot{x}^b = 0, \tag{2.5.6}$$

and raising c gives

$$\ddot{x}^c + \Gamma^c_{ab}\dot{x}^a\dot{x}^b = 0, \tag{2.5.7}$$

which are the equations of an affinely parametrised geodesic.

Those familiar with the calculus of variations or Lagrangian mechanics will know that the Euler–Lagrange equations give the solution to the problem of finding the curve (with fixed endpoints) which extremises the integral $\int_{u_1}^{u_2} L(\dot{x}^c, x^c)\,\mathrm{d}u$. While there is some connection with the characterisation of a geodesic as an extremal of length, it should be noted that the integral involved does not give the length of the curve. For reasons stated earlier, we shall not pursue this approach any further, but simply regard the Euler–Lagrange equations as a useful device for generating the geodesic equations and the connection coefficients which may be extracted from them.

Demonstrating the equivalence of the geodesic and the Euler–Lagrange equations allows us to make a useful observation. If g_{ab} does not depend on some particular coordinate x^d, say, then equation (2.5.5) shows that

$$\frac{\mathrm{d}}{\mathrm{d}u}\left(\frac{\partial L}{\partial \dot{x}^d}\right) = 0,$$

which implies that $\partial L/\partial \dot{x}^d$ is constant along an affinely parametrised geodesic. But $\partial L/\partial \dot{x}^d = g_{db}\dot{x}^b$, so we then have that $p_d \equiv g_{db}\dot{x}^b$

is constant along an affinely parametrised geodesic. The situation is exactly the same as in Lagrangian mechanics where, if the Lagrangian does not contain a particular generalised coordinate, then the corresponding generalised momentum is conserved, and borrowing a term from mechanics we call a coordinate which is absent from g_{ab} *cyclic* or *ignorable* [6]. Being able to say that p_d = constant whenever x^d is cyclic gives us immediate integrals of the geodesic equations, which help with their solution. An example should make some of these ideas clear.

Example.   The Robertson–Walker line element is used in cosmology (see Chapter 6). It is defined by

$$g_{\mu\nu}\mathrm{d}x^\mu\mathrm{d}x^\nu \equiv \mathrm{d}t^2 - [R(t)]^2[(1-kr^2)^{-1}\,\mathrm{d}r^2 + r^2\,\mathrm{d}\theta^2 + r^2\sin^2\theta\,\mathrm{d}\phi^2],$$

where μ, $\nu = 0$, 1, 2, 3 (our usual notation for spacetimes), k is a constant, and $x^0 \equiv t$, $x^1 \equiv r$, $x^2 \equiv \theta$, $x^3 \equiv \phi$.

So the Lagrangian is

$$L(\dot{x}^\sigma, x^\sigma) \equiv \tfrac{1}{2}\{\dot{t}^2 - [R(t)]^2[(1-kr^2)^{-1}\dot{r}^2 + r^2\dot{\theta}^2 + r^2\sin^2\theta\dot{\phi}^2]\},$$

where dots denote differentiation with respect to an affine parameter u. Partial differentiation gives:

$$\partial L/\partial\dot{t} = \dot{t},$$
$$\partial L/\partial\dot{r} = -R^2(1-kr^2)^{-1}\dot{r},$$
$$\partial L/\partial\dot{\theta} = -R^2 r^2\dot{\theta},$$
$$\partial L/\partial\dot{\phi} = -R^2 r^2\sin^2\theta\dot{\phi},$$
$$\partial L/\partial t = -RR'[(1-kr^2)^{-1}\dot{r}^2 + r^2\dot{\theta}^2 + r^2\sin^2\theta\dot{\phi}^2]$$

(where $R' = \mathrm{d}R/\mathrm{d}t$),

$$\partial L/\partial r = -R^2(1-kr^2)^{-1}kr\dot{r}^2 - R^2 r\dot{\theta}^2 - R^2 r\sin^2\theta\dot{\phi}^2,$$
$$\partial L/\partial\theta = -R^2 r^2\sin\theta\cos\theta\dot{\phi}^2,$$
$$\partial L/\partial\phi = 0.$$

Substitution of these derivatives in the Euler–Lagrange equations (2.5.5) gives:

$$\ddot{t} + RR'[(1-kr^2)^{-1}\dot{r}^2 + r^2\dot{\theta}^2 + r^2\sin^2\theta\dot{\phi}^2] = 0,$$
$$-R^2(1-kr^2)^{-1}\ddot{r} - 2RR'(1-kr^2)^{-1}\dot{t}\dot{r} - R^2(1-kr^2)^{-2}kr\dot{r}^2$$
$$+ R^2 r\dot{\theta}^2 + R^2 r\sin^2\theta\dot{\phi}^2 = 0,$$

$$-R^2r^2\ddot\theta - 2RR'r^2\dot t\dot\theta - 2R^2r\dot r\dot\theta + R^2r^2\sin\theta\cos\theta\dot\phi^2 = 0,$$
$$-R^2r^2\sin^2\theta\ddot\phi - 2RR'r^2\sin^2\theta\dot t\dot\phi - 2R^2r\sin^2\theta\dot r\dot\phi$$
$$-2R^2r^2\sin\theta\cos\theta\dot\theta\dot\phi = 0.$$

The above comprise the covariant version (2.5.6) of the geodesic equations. Obtaining the contravariant form (2.5.7) is in this case a simple matter of division (because $[g_{\mu\nu}]$ is diagonal):

$$\ddot t + RR'[(1-kr^2)^{-1}\dot r^2 + r^2\dot\theta^2 + r^2\sin^2\theta\dot\phi^2] = 0,$$
$$\ddot r + 2R'R^{-1}\dot t\dot r + kr(1-kr^2)^{-1}\dot r^2 - r(1-kr^2)\dot\theta^2 - r(1-kr^2)\sin^2\theta\dot\phi^2 = 0,$$
$$\ddot\theta + 2R'R^{-1}\dot t\dot\theta + 2r^{-1}\dot r\dot\theta - \sin\theta\cos\theta\dot\phi^2 = 0,$$
$$\ddot\phi + 2R'R^{-1}\dot t\dot\phi + 2r^{-1}\dot r\dot\phi + 2\cot\theta\dot\theta\dot\phi = 0.$$

Comparing these with equations (2.5.7) allows us to pick out the connection coefficients. These are zero except for the following:

$$\Gamma^0_{11} = RR'/(1-kr^2), \quad \Gamma^0_{22} = RR'r^2, \qquad \Gamma^0_{33} = RR'r^2\sin^2\theta,$$
$$\Gamma^1_{01} = R'/R, \qquad\qquad \Gamma^1_{11} = kr/(1-kr^2), \quad \Gamma^1_{22} = -r(1-kr^2),$$
$$\Gamma^1_{33} = -r(1-kr^2)\sin^2\theta,$$
$$\Gamma^2_{02} = R'/R, \qquad\qquad \Gamma^2_{12} = 1/r, \qquad\qquad \Gamma^2_{33} = -\sin\theta\cos\theta,$$
$$\Gamma^3_{03} = R'/R, \qquad\qquad \Gamma^3_{13} = 1/r, \qquad\qquad \Gamma^3_{23} = \cot\theta.$$

Note, for example, that in the second geodesic equation $2R'R^{-1}\dot t\dot r$ includes two terms of the sum $\Gamma^1_{\mu\nu}\dot x^\mu\dot x^\nu$, namely $\Gamma^1_{01}\dot x^0\dot x^1$ and $\Gamma^1_{10}\dot x^1\dot x^0$, and one must remember to halve the multipliers of the cross-terms $\dot x^\mu\dot x^\nu$ ($\mu \neq \nu$) when extracting the connection coefficients from the geodesic equations.

Note also that in this example ϕ is a cyclic coordinate, so one may say immediately that $\partial L/\partial\dot\phi$ is constant, i.e. $R^2r^2\sin^2\theta\dot\phi = A$, A = constant. Differentiation with respect to u results in the last geodesic equation, showing that we do indeed have an integral.

Exercises 2.5

1. Show that equation (2.5.3) is equivalent to

$$d^2u/dv^2 - h(v)\,du/dv = 0.$$

Hence obtain u explicitly as a function of v involving integrals of $h(v)$.

2. If a vector is transported parallelly along a geodesic, show that the angle between the vector and the tangent vector to the geodesic remains constant.

2.6   The spacetime of general relativity

The spacetime of special relativity is discussed in the Appendix. It is a 4-dimensional Riemannian manifold with the property that there exist global coordinate systems in which the metric tensor takes the form

$$[\eta_{\mu\nu}] \equiv \begin{bmatrix} 1 & 0 & 0 & 0 \\ 0 & -1 & 0 & 0 \\ 0 & 0 & -1 & 0 \\ 0 & 0 & 0 & -1 \end{bmatrix}$$

and we call such coordinate systems *inertial* or *cartesian*. We shall use x^μ to label points in spacetime, where the Greek suffixes, μ, ν, etc., have the range 0, 1, 2, 3, there being a certain convenience in counting from zero rather than one. As is customary in relativity we shall frequently refer to a point in spacetime as an *event*. Cartesian coordinates are related to the more familiar coordinates t, x, y, z of special relativity by $x^0 \equiv ct$, $x^1 \equiv x$, $x^2 \equiv y$, $x^3 \equiv z$, c being the velocity of light. We may, of course, use non-cartesian coordinates, where the metric tensor $g_{\mu\nu} \neq \eta_{\mu\nu}$, but the essential feature of the spacetime of special relativity is that we may always introduce a cartesian coordinate system about any point, so that $g_{\mu\nu} = \eta_{\mu\nu}$, and this coordinate system is global in the sense that its coordinate neighbourhood is the whole of spacetime.

One of our guiding requirements for the spacetime of general relativity is that locally it should be like the spacetime of special relativity. We therefore assume that it is a 4-dimensional Riemannian manifold with the property that about any point there exists a coordinate system in which the metric tensor field $g_{\mu\nu}$ is approximately $\eta_{\mu\nu}$. Note that we do *not* assert the existence of coordinate systems in which $g_{\mu\nu} = \eta_{\mu\nu}$ exactly, and this is the essential difference between the spacetimes of general and special relativity.

Making use of Exercise 2.4.3, we can construct a coordinate

system about any point P of general-relativistic spacetime in which $(\Gamma^\mu_{\nu\sigma})_P = 0$, and $(x^\mu)_P = (0, 0, 0, 0)$. Thus $(\partial_\sigma g_{\mu\nu})_P = 0$, and so for points near to P, where the coordinates x^μ are small, Taylor's theorem gives

$$g_{\mu\nu} \simeq (g_{\mu\nu})_P + \tfrac{1}{2}(\partial_\alpha \partial_\beta g_{\mu\nu})_P x^\alpha x^\beta. \qquad (2.6.1)$$

A linear transformation of coordinates given by $x^{\mu'} = A^{\mu'}_\nu x^\nu$, where $A^{\mu'}_\nu$ are constants and $[A^{\mu'}_\nu]$ is non-singular, may be used to reduce the matrix $[(g_{\mu\nu})_P]$ to diagonal form, with ± 1 as the diagonal entries, and with all the plus ones preceding the minus ones down the diagonal [7]. Moreover a linear transformation of coordinates results in $(\partial_{\sigma'} g_{\mu'\nu'})_P = 0$ and $(x^{\mu'})_P = (0, 0, 0, 0)$, so in the new coordinate system an equation analogous to equation (2.6.1) holds, but with $[(g_{\mu'\nu'})_P]$ diagonal. The requirement that the spacetime of general relativity be locally like that of special relativity then implies that this diagonal matrix is $[\eta_{\mu\nu}]$. So we have constructed a coordinate system about P in which for points near P we have, on dropping the prime,

$$g_{\mu\nu} \simeq \eta_{\mu\nu} + \tfrac{1}{2}(\partial_\alpha \partial_\beta g_{\mu\nu})_P x^\alpha x^\beta, \qquad (2.6.2)$$

and this approximation is valid for small x^μ.

If we are sufficiently close to P for the second term on the right of equation (2.6.2) to be neglected, we have a coordinate system in which $g_{\mu\nu} = \eta_{\mu\nu}$ approximately, and the extent of the region in which this approximation is valid will depend on the sizes of the second derivatives $(\partial_\alpha \partial_\beta g_{\mu\nu})_P$, and also on the accuracy of our measuring procedures. In this way we obtain *local inertial* or *local cartesian* coordinate systems about a point. It should be stressed that in special relativity we have *global* cartesian coordinate systems, where $g_{\mu\nu} = \eta_{\mu\nu}$ *exactly*, whereas in general relativity we have only *local* cartesian coordinate systems of limited extent, where $g_{\mu\nu} = \eta_{\mu\nu}$ *approximately*. We distinguish the two by saying that the spacetime of special relativity is *flat*, while that of general relativity is *curved*. The above discussion shows that the departure from flatness is connected with the non-vanishing of the second derivatives $\partial_\alpha \partial_\beta g_{\mu\nu}$, and we shall see the significance of this in Chapter 3, when we give a more formal definition of flatness in terms of the curvature tensor.

The purpose of the above discussion was to show, by introducing local cartesian coordinates, the sense in which the spacetime of general relativity is locally like that of special relativity. However, it is not sensible to work in terms of local cartesian coordinates as these involve approximations which amount to neglecting gravity, nor is it often convenient, since more suitable coordinates may be defined in a natural way. We therefore use general coordinates, and formulate things in ways which are valid in any coordinate system.

Another feature of the above discussion is that it gives us a means of generalising to general relativity results which are valid in special relativity. For example, it is shown in the Appendix that in a cartesian coordinate system of special-relativistic spacetime, Maxwell's equations may be written in the form

$$\left.\begin{array}{r} F^{\mu\nu}{}_{,\nu} = \mu_0 j^{\mu}, \\ F_{\mu\nu,\sigma} + F_{\nu\sigma,\mu} + F_{\sigma\mu,\nu} = 0, \end{array}\right\} \qquad (2.6.3)$$

where a comma denotes partial differentiation. We may adopt

$$\left.\begin{array}{r} F^{\mu\nu}{}_{;\nu} = \mu_0 j^{\mu}, \\ F_{\mu\nu;\sigma} + F_{\nu\sigma;\mu} + F_{\sigma\mu;\nu} = 0, \end{array}\right\} \qquad (2.6.4)$$

where now a semicolon denotes covariant differentiation, as the general-relativistic version of these, for in a local cartesian coordinate system (where $g_{\mu\nu} = \eta_{\mu\nu}$ approximately, and we can neglect $\Gamma^{\mu}_{\nu\sigma}$) equations (2.6.4) reduce to (2.6.3). There are really two points to note here. The first is that if any physical quantity can be defined as a cartesian tensor in special relativity, then we can give its definition in general relativity by defining it in exactly the same way in a local cartesian coordinate system; its components in any other coordinate system are then given by the usual transformation formulae (1.4.2). Given this first point, the second is that any cartesian tensor equation valid in special relativity may be converted to an equation valid in general relativity in any coordinate system, simply by replacing partial differentiation with respect to coordinates by covariant differentiation, total derivatives along curves by absolute derivatives, and $\eta_{\mu\nu}$ by $g_{\mu\nu}$. (Compare remarks made in the Introduction.)

As an example of this, consider the path of a particle (with mass) in special relativity. Its world velocity is $u^{\mu} \equiv dx^{\mu}/d\tau$ (see Section A.5), where the proper time τ for the particle is defined by (see Section A.0)

$$c^2 \, d\tau^2 \equiv \eta_{\mu\nu} \, dx^{\mu} \, dx^{\nu}.$$

Its equation of motion is then (equation (A.6.8))

$$dp^\mu/d\tau = f^\mu,$$

where $p^\mu \equiv mu^\mu$, m being the proper mass of the particle and f^μ the 4-force acting on it. The generalisation of these ideas to general relativity gives $u^\mu \equiv dx^\mu/d\tau$ as the definition of the world velocity of the particle, where now the proper time τ is defined by

$$c^2 d\tau^2 \equiv g_{\mu\nu} dx^\mu dx^\nu, \qquad (2.6.5)$$

and

$$Dp^\mu/d\tau = f^\mu \qquad (2.6.6)$$

as the equation of motion, where $p^\mu \equiv mu^\mu$, and the definitions of m and f^μ are taken over from special relativity as explained above. Moreover, these equations are valid in any coordinate system.

As in special relativity we assume that a "good" clock measures its own proper time. In particular, if the particle is a pulsating atom, the proper time interval between events on the atom's path where successive pulses occur is constant.

In the case of a free particle for which $f^\mu = 0$, equation (2.6.6) reduces to $D(dx^\mu/d\tau)/d\tau = 0$, or

$$\frac{d^2 x^\mu}{d\tau^2} + \Gamma^\mu_{\nu\sigma} \frac{dx^\nu}{d\tau} \frac{dx^\sigma}{d\tau} = 0. \qquad (2.6.7)$$

From this we conclude that the path of a free particle is a geodesic in spacetime, and that the proper time experienced by the particle is an affine parameter along it. This result is often stated as an explicit postulate of general relativity (the *geodesic postulate*), but it emerges here as a natural consequence of the way in which we generalise special-relativistic concepts. It is a perfectly natural generalisation, for the path of a free particle in the flat spacetime of special relativity is a straight line and this generalises to a geodesic in curved spacetime.

The path of a photon (or any other zero-rest-mass particle) in the spacetime of special relativity is also a straight line, and this also generalises to a geodesic in curved spacetime. However, there is no change in proper time along the path of a photon, so τ cannot be used as a parameter. But we can still use an affine parameter u, so that the analogue of equation (2.6.7) for a photon is

$$\frac{\mathrm{d}^2 x^\mu}{\mathrm{d}u^2} + \Gamma^\mu_{\nu\sigma} \frac{\mathrm{d}x^\nu}{\mathrm{d}u} \frac{\mathrm{d}x^\sigma}{\mathrm{d}u} = 0 \quad . \tag{2.6.8}$$

The fact that the photon's speed is c finds expression as

$$g_{\mu\nu} \frac{\mathrm{d}x^\mu}{\mathrm{d}u} \frac{\mathrm{d}x^\nu}{\mathrm{d}u} = 0 \quad , \tag{2.6.9}$$

which generalises the relation $\eta_{\mu\nu} \dfrac{\mathrm{d}x^\mu}{\mathrm{d}u} \dfrac{\mathrm{d}x^\nu}{\mathrm{d}u} = 0$ (or $c^2\,\mathrm{d}t^2 - \mathrm{d}x^2 - \mathrm{d}y^2 - \mathrm{d}z^2 = 0$) of special relativity.

Another concept of special relativity which has a natural generalisation is that of a vector being timelike, null or spacelike. It is clear that the appropriate definition is that the vector λ^μ is

$$
\begin{matrix} \textit{timelike} \\ \textit{null} \\ \textit{spacelike} \end{matrix}
\quad \text{if} \quad g_{\mu\nu}\lambda^\mu\lambda^\nu
\quad \begin{cases} >0 \\ =0 \\ <0 \end{cases}
$$

One should note that at any point of spacetime the null cone of vectors given by $g_{\mu\nu}\lambda^\mu\lambda^\nu = 0$ lies in the tangent space at that point and not in the manifold. The fact is not readily appreciated in the flat spacetime of special relativity, because its basic linear structure allows one to regard the tangent space at each point as being embedded in the spacetime.

At any point on the path of a particle (with mass) its world velocity is a tangent vector to the path, and equation (2.6.5) tells us that this tangent vector is timelike. We therefore say that a particle with mass follows a *timelike path* through *spacetime*, and in particular a free particle follows a *timelike geodesic*. A photon, however, follows a *null geodesic*, as equation (2.6.9) tells us that the tangent vectors to its path are null. Spacelike paths and spacelike geodesics may also be defined, but these have no physical significance [8].

In moving from the flat spacetime of special relativity to the curved spacetime of general relativity we hope somehow to incorporate the effects of gravity, and the point of view we are adopting is that gravity is not a force, and that gravitational effects may be explained in terms of the curvature of spacetime. It should therefore be understood that by free particles we mean particles moving under gravity alone. Comparing equation (2.6.7) with its special-relativistic analogue $d^2x^\mu/d\tau^2 = 0$ indicates that the connection coefficients play an important role in explaining gravitational effects. Since these are given by derivatives of the metric tensor field, we see that it is this tensor field which, in a sense, carries the gravitational content of spacetime. For the moment we shall take the metric tensor field as given, and postpone until Chapter 3 the question of how it is determined by the distribution of matter and energy in spacetime. In the rest of this chapter we take a closer look at equations (2.6.6) and (2.6.7), and relate them to some familiar Newtonian ideas.

Exercises 2.6

1. Show that if at a point of spacetime the non-zero vector λ^μ is orthogonal to:
   (a) a timelike vector t^μ, then λ^μ is spacelike;
   (b) a null vector n^μ, then λ^μ is either spacelike or proportional to n^μ;
   (c) a spacelike vector, then λ^μ may be either timelike, null or spacelike.

2.7   Newton's laws of motion

Newton's first law that "every body perseveres in its state of rest, or of uniform motion in a right [straight] line, unless it is compelled to change that state by forces impressed thereon" clearly has its counterpart in the statement that "every particle follows a geodesic in spacetime" [9]. Indeed, in a local inertial coordinate system where we may neglect the $\Gamma^\mu_{\nu\sigma}$, the geodesic equation reduces to $d^2x^\mu/d\tau^2 = 0$. For non-relativistic speeds $d\tau/dt$ is approximately 1, so the geodesic equation yields $d^2x^i/dt^2 = 0$, $i = 1, 2, 3$, the familiar Newtonian equation of motion of a free particle.

Newton's second law that "the alteration of motion is ever proportional to the motive force impressed; and is made in the right line in which that force is impressed" is usually rendered as the 3-vector equation

$$d\mathbf{p}/dt = \mathbf{F},$$

Table 2.1.   Newton's laws and their general-relativistic counterparts.

Newton	Einstein
Free particles move in straight lines through space	Free particles follow geodesics through spacetime
$$\mathbf{F} = m\frac{d^2\mathbf{x}}{dt^2}$$	$$f^\mu = m\left(\frac{d^2x^\mu}{d\tau^2} + \Gamma^\mu_{\nu\sigma}\frac{dx^\nu}{d\tau}\frac{dx^\sigma}{d\tau}\right)$$
To every action there is always opposed an equal reaction	Interaction of two large bodies is not simple; each body produces curvature of spacetime

where **p** is the momentum and **F** the applied force. This clearly has its counterpart in equation (2.6.6).

Newton's third law that "to every action there is always opposed an equal reaction: or the mutual actions of two bodies upon each other are always equal, and directed to contrary parts" loses its meaning in general relativity, at least as far as gravitational interaction is concerned. The reason is that we no longer regard gravity as a force, and we have effectively adopted a *field* viewpoint. The gravitational field of a body is contained in the curvature of spacetime that it produces. A particle responds to the field by following a geodesic in the curved spacetime of the gravitating body. It should be noted that this viewpoint ignores any curvature produced by the particle following the geodesic. That is, the particle is a *test particle*, and there is no question of its having any effect on the body producing the gravitational field. The interaction of two extended bodies may be discussed in general relativity via the field equations of the theory, but this is not a simple business, and approximation methods must be used.

2.8   Gravitational potential and the geodesic

Suppose we have a coordinate system in which the metric tensor field is given by

$$g_{\mu\nu} \equiv \eta_{\mu\nu} + h_{\mu\nu}, \tag{2.8.1}$$

where the $h_{\mu\nu}$ are small, but not so small that they may be neglected. In the language of Section 2.0, our coordinate system is bad, but not too bad. Our aim in this section is to obtain a

Newtonian approximation to the geodesic equation given by the metric tensor (2.8.1) valid for a particle whose velocity components $\mathrm{d}x^i/\mathrm{d}t$ ($i = 1, 2, 3$) are small compared with c. We shall assume that the gravitational field, as expressed by $h_{\mu\nu}$, is *quasi-static* in the sense that $\partial_0 h_{\mu\nu} \equiv c^{-1}\partial h_{\mu\nu}/\partial t$ is negligible when compared with $\partial_i h_{\mu\nu}$.

If instead of the proper time τ we use the coordinate time t (defined by $x^0 \equiv ct$) as a parameter, then the geodesic equation giving the path of a free particle has the form

$$\frac{\mathrm{d}^2 x^\mu}{\mathrm{d}t^2} + \Gamma^\mu_{\nu\sigma}\frac{\mathrm{d}x^\nu}{\mathrm{d}t}\frac{\mathrm{d}x^\sigma}{\mathrm{d}t} = h(t)\frac{\mathrm{d}x^\mu}{\mathrm{d}t}, \tag{2.8.2}$$

where

$$h(t) \equiv -\frac{\mathrm{d}^2 t}{\mathrm{d}\tau^2}\bigg/\left(\frac{\mathrm{d}t}{\mathrm{d}\tau}\right)^2 = \frac{\mathrm{d}^2\tau}{\mathrm{d}t^2}\bigg/\frac{\mathrm{d}\tau}{\mathrm{d}t}. \tag{2.8.3}$$

This follows from equation (2.5.2). On dividing by c^2, the spatial part of equation (2.8.2) may be written

$$\frac{1}{c^2}\frac{\mathrm{d}^2 x^i}{\mathrm{d}t^2} + \Gamma^i_{00} + 2\Gamma^i_{0j}\left(\frac{1}{c}\frac{\mathrm{d}x^j}{\mathrm{d}t}\right) + \Gamma^i_{jk}\left(\frac{1}{c}\frac{\mathrm{d}x^j}{\mathrm{d}t}\right)\left(\frac{1}{c}\frac{\mathrm{d}x^k}{\mathrm{d}t}\right)$$
$$= \frac{1}{c}h(t)\left(\frac{1}{c}\frac{\mathrm{d}x^i}{\mathrm{d}t}\right), \quad (2.8.4)$$

and the last term on the left is clearly negligible.

If we put $h^{\mu\nu} \equiv \eta^{\mu\sigma}\eta^{\nu\rho}h_{\sigma\rho}$, then a short calculation shows that, to first order in the small quantities $h_{\mu\nu}$ and $h^{\mu\nu}$,

$$g^{\mu\nu} = \eta^{\mu\nu} - h^{\mu\nu} \quad \text{and} \quad \Gamma^\mu_{\nu\sigma} = \tfrac{1}{2}\eta^{\mu\rho}(\partial_\nu h_{\sigma\rho} + \partial_\sigma h_{\nu\rho} - \partial_\rho h_{\nu\sigma}). \tag{2.8.5}$$

So to first order,

$$\Gamma^i_{00} = \tfrac{1}{2}\eta^{i\rho}(\partial_0 h_{0\rho} + \partial_0 h_{0\rho} - \partial_\rho h_{00})$$
$$= -\tfrac{1}{2}\eta^{ij}\partial_j h_{00} = \tfrac{1}{2}\delta^{ij}\partial_j h_{00},$$

on neglecting $\partial_0 h_{\mu\nu}$ in comparison with $\partial_i h_{\mu\nu}$. Also to first order,

$$\Gamma^i_{0j} = \tfrac{1}{2}\eta^{i\rho}(\partial_0 h_{j\rho} + \partial_j h_{0\rho} - \partial_\rho h_{0j})$$
$$= -\tfrac{1}{2}\delta^{ik}(\partial_j h_{0k} - \partial_k h_{0j}),$$

again on neglecting $\partial_0 h_{\mu\nu}$.

We have now approximated all the terms on the left-hand side of equation (2.8.4), and there remains the right-hand side to deal with. Working to the same level of approximation as above, and neglecting squares and products of $c^{-1} \, dx^i/dt$, we find from

$$\left(\frac{d\tau}{dt}\right)^2 = \frac{1}{c^2} \, g_{\mu\nu} \frac{dx^\mu}{dt} \frac{dx^\nu}{dt}$$

that

$$d\tau/dt = (1 + h_{00})^{1/2} = 1 + \tfrac{1}{2} h_{00}, \qquad (2.8.6)$$

so

$$d^2\tau/dt^2 = \tfrac{1}{2} c h_{00,0},$$

and

$$\frac{1}{c} \, h(t) = \tfrac{1}{2} h_{00,0} (1 - \tfrac{1}{2} h_{00}) = \tfrac{1}{2} h_{00,0},$$

from equation (2.8.3).

It follows that the right-hand side of equation (2.8.4) is negligible, and our approximation gives

$$\frac{1}{c^2} \frac{d^2 x^i}{dt^2} + \tfrac{1}{2} \delta^{ij} \partial_j h_{00} - \delta^{ik} (\partial_j h_{0k} - \partial_k h_{0j}) \frac{1}{c} \frac{dx^j}{dt} = 0.$$

Introducing the mass m of the particle and rearranging gives

$$m \frac{d^2 x^i}{dt^2} = -m \delta^{ij} \partial_j (\tfrac{1}{2} c^2 h_{00}) + mc \delta^{ik} (\partial_j h_{0k} - \partial_k h_{0j}) \frac{dx^j}{dt}. \quad (2.8.7)$$

Let us now interpret this in Newtonian terms. Recall that when discussing the allegory of the apple in Section 2.0, we stated that if we thought a bad coordinate system were a good one, a free particle following a geodesic would appear to be acted on by a force, which is conventionally referred to as gravity. The left-hand side is mass $\times$ acceleration, so the right-hand side is the 'gravitational force' on the particle. The first term on the right is the force $-m\nabla V$ arising from a potential V given by $V \equiv \tfrac{1}{2} c^2 h_{00}$, while the second term on the right is velocity-dependent and clearly smacks of rotation [10]. This is not surprising, for the principle of equivalence asserts that the forces of acceleration, such as the velocity-dependent Coriolis force [11] which would arise from using a rotating reference system, are on the same footing as gravitational forces. If we agree to call a nearly inertial coordinate system in which $\partial_j h_{0k} - \partial_k h_{0j}$ is zero *non-rotating*, then we have for a slowly moving particle in a nearly inertial non-rotating coordinate system, in which the quasi-static condition holds, the approximation

$$\mathrm{d}^2x^i/\mathrm{d}t^2 = -\delta^{ij}\partial_j V, \tag{2.8.8}$$

where

$$V \equiv \tfrac{1}{2}c^2 h_{00} + \text{const.} \tag{2.8.9}$$

This is the Newtonian equation of motion for a particle moving in a gravitational field of potential V, provided we make the identification (2.8.9). This gives

$$g_{00} = 2V/c^2 + \text{const},$$

and if we choose the constant to be 1, then g_{00} reduces to its flat-spacetime value when $V = 0$. This gives

$$g_{00} = 1 + 2V/c^2 \tag{2.8.10}$$

as the relation between g_{00} and the Newtonian potential V in this approximation.

Exercises 2.8

1. Check the approximations (2.8.5) and (2.8.6).

2.9   Newton's law of universal gravitation

Newton's law of universal gravitation does not survive intact in general relativity, which is after all a new theory replacing the Newtonian theory. However, we should be able to recover it as an approximation.

The Schwarzschild solution is an exact solution of the field equations of general relativity, and it may be identified as representing the field produced by a massive body. This solution is derived in the next chapter (see Section 3.5), and its line element is

$$c^2\,\mathrm{d}\tau^2 = (1 - 2GM/rc^2)c^2\,\mathrm{d}t^2 - (1 - 2GM/rc^2)^{-1}\,\mathrm{d}r^2$$
$$- r^2\,\mathrm{d}\theta^2 - r^2\sin^2\theta\,\mathrm{d}\phi^2,$$

where M is the mass of the body and G the gravitational constant. For small values of GM/rc^2 this is close to the line element of flat spacetime in spherical polar coordinates, and r then behaves like radial distance. If we were to put

$$x^0 \equiv ct, \qquad x^1 \equiv r \sin \theta \cos \phi, \qquad x^2 \equiv r \sin \theta \sin \phi, \qquad x^3 \equiv r \cos \theta,$$

we would obtain a line element whose metric tensor had the form $g_{\mu\nu} = \eta_{\mu\nu} + h_{\mu\nu}$, where, for large values of rc^2/GM, the $h_{\mu\nu}$ are small and $g_{00} = 1 - 2GM/rc^2$. This gives $h_{00} = -2GM/rc^2$, and according to the results of the last section, a Newtonian potential $V = -GM/r$. The 3-vector form of equation (2.8.8) gives

$$m \, d^2\mathbf{r}/dt^2 = -m\nabla V = -GM \, mr^{-2}\hat{\mathbf{r}},$$

where $\mathbf{r} \equiv (x^1, x^2, x^3)$, m is the mass of the test particle, and $\hat{\mathbf{r}}$ is a unit vector in the direction of $\mathbf{r}$. The "force" on the test particle is in agreement with that given by Newton's law, and in this way the law is recovered as an approximation valid for large values of rc^2/GM and slowly-moving particles.

2.10   A rotating reference system

The principle of equivalence (see the Introduction) implies that the "fictitious" forces of accelerating coordinate systems are essentially in the same category as the "real" forces of gravity. Put another way, if the geodesic equation contains gravity in the $\Gamma^\mu_{\nu\sigma}$ it must also contain any accelerations which may have been built in by choice of coordinate system. In a curved spacetime it is not always easy, and often impossible, to sort these "forces" out, but in flat spacetime we have only the "fictitious" forces of acceleration and these should be included in the $\Gamma^\mu_{\nu\sigma}$. As an example of this, let us consider a rotating reference system in flat spacetime.

Starting with a non-rotating system K with coordinates (T, X, Y, Z) and line element

$$c^2 \, d\tau^2 = c^2 \, dT^2 - dX^2 - dY^2 - dZ^2, \tag{2.10.1}$$

let us define new coordinates (t, x, y, z) by (see Fig. 2.7)

$$\left.\begin{aligned}
T &\equiv t, \\
X &\equiv x \cos \omega t - y \sin \omega t, \\
Y &\equiv x \sin \omega t + y \cos \omega t, \\
Z &\equiv z.
\end{aligned}\right\} \tag{2.10.2}$$

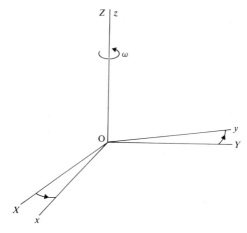

Fig. 2.7 Coordinate system $K'(x, y, z)$ rotating relative to the coordinate system $K(X, Y, Z)$.

Points given by x, y, z constant rotate with angular speed ω about the Z-axis of K, and this defines the rotating system K' (see Fig. 2.7). In terms of the new coordinates the line element is

$$c^2 \, d\tau^2 = [c^2 - \omega^2 (x^2 + y^2)] \, dt^2 + 2\omega y \, dx \, dt$$
$$- 2\omega x \, dy \, dt - dx^2 - dy^2 - dz^2, \quad (2.10.3)$$

and the geodesic equations are:

$$\ddot{t} = 0, \qquad (2.10.4a)$$
$$\ddot{x} - \omega^2 x \dot{t}^2 - 2\omega \dot{y} \dot{t} = 0, \qquad (2.10.4b)$$
$$\ddot{y} - \omega^2 y \dot{t}^2 + 2\omega \dot{x} \dot{t} = 0, \qquad (2.10.4c)$$
$$\ddot{z} = 0, \qquad (2.10.4d)$$

where dots denote differentiation with respect to proper time (see Exercises 2.10.1 and 2.10.2). These constitute the equation of motion of a free particle (with mass).

Equation (2.10.4a) implies that $dt/d\tau$ is constant, so the remaining equations may be written as

$$d^2x/dt^2 - \omega^2 x - 2\omega \, dy/dt = 0,$$
$$d^2y/dt^2 - \omega^2 y + 2\omega \, dx/dt = 0,$$
$$d^2z/dt^2 = 0.$$

Introducing the mass m of the particle and rearranging gives

$$m \, d^2x/dt^2 = m\omega^2 x + 2m\omega \, dy/dt,$$
$$m \, d^2y/dt^2 = m\omega^2 y - 2m\omega \, dx/dt, \qquad (2.10.5)$$
$$m \, d^2z/dt^2 = 0,$$

or, in 3-vector notation,

$$m \, d^2\mathbf{r}/dt^2 = -m\boldsymbol{\omega} \times (\boldsymbol{\omega} \times \mathbf{r}) - 2m\boldsymbol{\omega} \times (d\mathbf{r}/dt), \qquad (2.10.6)$$

where $\mathbf{r} \equiv (x, y, z)$ and $\boldsymbol{\omega} \equiv (0, 0, \omega)$.

A Newtonian observer at rest in the rotating system K' would interpret the left-hand side of equation (2.10.6) as mass × acceleration, and would therefore assert the existence of a "gravitational force" as given by the right-hand side. This "force" is, of course, the sum of the centrifugal force $-m\boldsymbol{\omega} \times (\boldsymbol{\omega} \times \mathbf{r})$ and the Coriolis force $-2m\boldsymbol{\omega} \times (d\mathbf{r}/dt)$ [11], and the geodesic equation does indeed include the forces of acceleration in the $\Gamma^{\mu}_{\nu\sigma}$.

In the above all calculations were exact, but we can relate things to the approximation methods of Section 2.8 by putting $x^0 \equiv ct$, $x^1 \equiv x$, $x^2 \equiv y$, $x^3 \equiv z$, and noting that the line element (2.10.3) then gives $g_{\mu\nu} = \eta_{\mu\nu} + h_{\mu\nu}$, where

$$[h_{\mu\nu}] \equiv \begin{bmatrix} -\omega^2(x^2+y^2)/c^2 & \omega y/c & -\omega x/c & 0 \\ \omega y/c & 0 & 0 & 0 \\ -\omega x/c & 0 & 0 & 0 \\ 0 & 0 & 0 & 0 \end{bmatrix}.$$

The $h_{\mu\nu}$ are small, provided we restrict ourselves to the region near the z-axis where $\omega^2(x^2+y^2)/c^2$ is small. Moreover, $\partial_0 h_{\mu\nu} = 0$, so the quasi-static condition is fulfilled. However, our system is rotating, so we must use the approximation (2.8.7) rather than (2.8.8). We see that

$$\tfrac{1}{2}c^2 h_{00} = -\tfrac{1}{2}\omega^2(x^2+y^2),$$

and a straightforward calculation (see Exercise 2.10.4) gives

$$[A^i_j] = \begin{bmatrix} 0 & 2\omega & 0 \\ -2\omega & 0 & 0 \\ 0 & 0 & 0 \end{bmatrix}, \qquad (2.10.7)$$

where $A_j^i \equiv c\delta^{ik}(\partial_j h_{0k} - \partial_k h_{0j})$. Hence the approximation (2.8.7) gives

$$m\, d^2x/dt^2 = m\omega^2 x + 2m\omega\, dy/dt,$$

$$m\, d^2y/dt^2 = m\omega^2 y - 2m\omega\, dx/dt,$$

$$m\, d^2z/dt^2 = 0.$$

These equations are identical with equations (2.10.5), and may be rearranged to exhibit the centrifugal and Coriolis forces, as before. The centrifugal force has its origin in the potential $V \equiv -\frac{1}{2}\omega^2(x^2 + y^2)$.

Finally, we may draw another conclusion from the transformed line element (2.10.3). A clock fixed in the rotating system K' will have x, y, z constant, so

$$c^2\, d\tau^2 = [c^2 - \omega^2(x^2 + y^2)]\, dt^2,$$

giving

$$d\tau = [1 - \omega^2(x^2 + y^2)/c^2]^{1/2}\, dt, \tag{2.10.8}$$

as the proper time interval recorded by the clock. This means that the clock in the rotating system K' runs slowly when compared with a clock in the inertial system K. Relative to K, the speed of the clock is $\omega(x^2 + y^2)^{1/2}$, so the result (2.10.8) is in agreement with equation (A.0.6) of the Appendix. Note that our clock and other material objects may only be placed at rest in the rotating system where $x^2 + y^2 < c^2/\omega^2$, for otherwise their speed (relative to K) exceeds c.

Exercises 2.10

1. Check the form of the line element (2.10.3).
2. Obtain the geodesic equations (2.10.4) in three different ways:
    (i) By using the Euler–Lagrange equations;
    (ii) By extracting $[g_{\mu\nu}]$ from the line element (2.10.3), inverting to find $[g^{\mu\nu}]$, and then calculating the $\Gamma_{\nu\sigma}^\mu$;
    (iii) By substituting for T, X, Y, Z in $\ddot{T} = \ddot{X} = \ddot{Y} = \ddot{Z} = 0$, using equations (2.10.2).
3. Cylindrical polar coordinates (ρ, ϕ, z) may be introduced into the rotating system K' by putting $x \equiv \rho \cos\phi$, $y \equiv \rho \sin\phi$. Show

that in terms of these the geodesic equations are

$$\ddot{t} = 0,$$
$$\ddot{\rho} - \rho\omega^2\dot{t}^2 - \rho\dot{\phi}^2 - 2\omega\rho\dot{\phi}\dot{t} = 0,$$
$$\ddot{\phi} + 2\rho^{-1}\dot{\rho}\dot{\phi} + 2\omega\rho^{-1}\dot{\rho}\dot{t} = 0,$$
$$\ddot{z} = 0,$$

so that corresponding to equations (2.10.5) one has

$$m\left[\frac{d^2\rho}{dt^2} - \rho\left(\frac{d\phi}{dt}\right)^2\right] = m\rho\omega^2 + 2m\omega\rho\frac{d\phi}{dt},$$

$$m\left(\rho\frac{d^2\phi}{dt^2} + 2\frac{d\rho}{dt}\frac{d\phi}{dt}\right) = -2m\omega\frac{d\rho}{dt},$$

$$m\frac{d^2z}{dt^2} = 0.$$

Interpret these in terms of the radial, transverse and axial components of acceleration, centrifugal and Coriolis forces.

4. Check that the matrix $[A^i_j]$ is as given by equation (2.10.7), and that the approximation (2.8.7) does give the equations (2.10.5).

Problems 2.

1. Show that in spherical polar coordinates the line element of three-dimensional euclidean space is

$$ds^2 = dr^2 + r^2\,d\theta^2 + r^2\sin^2\theta\,d\phi^2.$$

Deduce that the line element of:

(a) flat spacetime in spherical polar coordinates is

$$c^2\,d\tau^2 = c^2\,dt^2 - dr^2 - r^2\,d\theta^2 - r^2\sin^2\theta\,d\phi^2;$$

(b) a sphere of radius a embedded in three-dimensional euclidean space is

$$ds^2 = a^2(d\theta^2 + \sin^2\theta\,d\phi^2),$$

where θ, ϕ are the usual polar angles giving points on the sphere.

2. Let $\overrightarrow{OP} \equiv \delta_1 x^a$ and $\overrightarrow{OQ} \equiv \delta_2 x^a$ be small coordinate differences at a point O of a manifold. Transport $\overrightarrow{OP}$ parallelly along $\overrightarrow{OQ}$ to obtain $\overrightarrow{QP'}$, and $\overrightarrow{OQ}$ parallelly along $\overrightarrow{OP}$ to obtain $\overrightarrow{PQ'}$ (see Fig. 2.8).

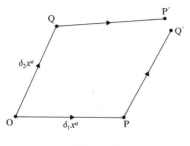

Fig. 2.8

Show that in general $P' = Q'$ (to second order in $\delta_1 x^a$, $\delta_2 x^a$) if and only if the connection is symmetric.

(This shows that 'infinitesimal' parallelograms exist if and only if the connection is symmetric.)

3. Show that in general, for a scalar field ϕ, $\phi_{;ab} = \phi_{;ba}$ if and only if the connection is symmetric.

4. The *curl* of a covariant vector field λ_a is the skew-symmetric tensor A_{ab} defined by

$$A_{ab} \equiv \lambda_{a;b} - \lambda_{b;a}.$$

Show that $A_{ab} = \lambda_{a,b} - \lambda_{b,a}$, provided the connection is symmetric.

5. If A_{ab} is a skew-symmetric type $(0, 2)$ tensor field, prove that

$$B_{abc} \equiv A_{ab,c} + A_{bc,a} + A_{ca,b}$$

are the components of a type $(0, 3)$ tensor field.

(Hint: Put $A_{ab,c} = A_{ab;c} + \Gamma^d_{ac} A_{db} + \Gamma^d_{bc} A_{ad}$, where the Γ^a_{bc} are the coefficients of a symmetric connection.)

6. Assuming that the geodesics on a sphere are great circles, use the result of Exercise 2.5.2 to show that, in general, parallel transport is path-dependent.

7. The line element of a static spherically symmetric spacetime is $c^2 \, d\tau^2 = A(r) \, dt^2 - B(r) \, dr^2 - r^2 \, d\theta^2 - r^2 \sin^2 \theta \, d\phi^2$. Use the Euler–Lagrange equations to obtain the geodesic equations, and hence show that the only non-vanishing connection coefficients are:

$\Gamma^0_{01} = A'/2A$, $\qquad$ $\Gamma^1_{00} = A'/2B$, $\qquad$ $\Gamma^1_{11} = B'/2B$,

$\Gamma^1_{22} = -r/B$, $\qquad$ $\Gamma^1_{33} = -(r \sin^2 \theta)/B$, $\quad$ $\Gamma^2_{12} = 1/r$,

$\Gamma^2_{33} = -\sin \theta \cos \theta$, $\qquad$ $\Gamma^3_{13} = 1/r$, $\qquad$ $\Gamma^3_{23} = \cot \theta$, where primes denote derivatives with respect to r, and

$$x^0 \equiv t, \quad x^1 \equiv r, \quad x^2 \equiv \theta, \quad x^3 \equiv \phi.$$

8. One can conceive of an observer in a swivel chair located above the Sun, looking down on the plane of the Earth's orbit. If the chair rotates at the rate of one revolution a year, then to the observer the Earth appears stationary. If for some reason all heavenly bodies other than the Earth and the Sun are invisible, how does the observer explain why the Earth does not collapse in towards the Sun, there being no detectable orbit?

Notes

1. Misner, Thorne and Wheeler, 1973, §1.1.
2. Compare remarks made in the Introduction.
3. Apostol, 1974, §§2.6, 3.2, 3.3.
4. See, for example, Apostol, 1974, §13.3.
5. Our definition of a tangent vector makes use of N-tuples related to a coordinate system, and in this respect is somewhat unsatisfactory. However, we believe that it has advantages of simplicity when compared with more sophisticated definitions, such as defining it as an equivalence class of curves or as a directional derivative.
6. See, for example, Goldstein, 1959, §2–6.
7. See, for example, Birkhoff and MacLane, 1977, §9.9.
8. Unless one believes in tachyons.
9. The versions of Newton's laws quoted here are from Andrew Motte's translation (London, 1729) of Newton's *Principia*.
10. Most authors assume that the derivatives $\partial_i h_{\mu\nu}$ are small along with the $h_{\mu\nu}$, and therefore do not obtain these velocity-dependent rotational terms. However, the fact that the $h_{\mu\nu}$ are small does not mean that their derivatives are also small (see Section 2.10).
11. See, for example, Goldstein, 1959, §4–9.

Field equations and curvature

3.0 Introduction

The main purpose of this chapter is to establish the field equations of general relativity, which couple the gravitational field (contained in the curvature of spacetime) with its sources. We start by discussing a tensor which effectively and concisely describes the sources, and follow that with a discussion of curvature, then bring these together in the field equations.

The field equation of Newton's theory is Poisson's equation, and we use it as a guide in constructing the field equations of general relativity, which, as we shall see, then yield Poisson's equation as an approximation.

The chapter finishes with an exact solution of the field equations representing the gravitational field of spherically symmetric massive body. This solution forms the basis of our discussions in Chapter 4.

3.1 The stress tensor and fluid motion

Except at the very end, where we take the step to curved spacetime, our discussion in this section takes place in flat spacetime, where we use inertial coordinate systems. We shall be dealing with 4-vectors and 3-vectors, and we shall use bold-faced type for the latter only. With some abuse of notation, we shall write

$$\lambda^{\mu} \equiv (\lambda^0, \lambda^1, \lambda^2, \lambda^3) \equiv (\lambda^0, \boldsymbol{\lambda}).$$

We start by considering a particle, and some quantities we make use of are:

$m \equiv$ rest or proper mass of a particle [1],

$t \equiv$ coordinate time,

$\tau \equiv$ proper time,

$\gamma \equiv dt/d\tau = (1 - v^2/c^2)^{-1/2}$, where v is the particle's speed,
$E \equiv \gamma mc^2 \equiv$ energy of particle,
$u^\mu \equiv dx^\mu/d\tau \equiv$ world velocity,
$v^\mu \equiv dx^\mu/dt = u^\mu/\gamma \equiv$ coordinate velocity,
$p^\mu \equiv mu^\mu \equiv$ 4-momentum of particle.

So in our notation, $v^\mu \equiv (c, \mathbf{v})$, where $\mathbf{v}$ is the particle's 3-velocity, so that v occurring in the formula for γ is $|\mathbf{v}|$ (see Appendix for details). Of the quantities listed above, only m and τ are scalars, and only u^μ and p^μ are vectors.

A stationary particle situated at the point with position vector $\mathbf{x}_0$ has

$$u^\mu \equiv dx^\mu/d\tau = d(c\tau, \mathbf{x}_0)/d\tau = (c, \mathbf{0}),$$

and

$$p^\mu = m(c, \mathbf{0}).$$

The zeroth component of p^μ is in this case the *rest energy* of the particle (up to a factor c). For a moving particle we have

$$p^\mu \equiv mu^\mu = \gamma mv^\mu = (\gamma mc, \gamma m\mathbf{v}) = (E/c, \mathbf{p}). \tag{3.1.1}$$

Equation (3.1.1) emphasises the fact that in relativity, energy and momentum are the temporal and spatial parts of a single 4-vector p^μ. They always maintain this distinction, even after (Lorentz) transformations, just as the 4-vector $x^\mu \equiv (ct, \mathbf{x})$ is split distinctly into two parts, time and position, with time being always the zeroth component.

Let us now pass to a continuous distribution of matter, and for simplicity we shall take it to be a perfect fluid, which is characterised by two scalar fields, namely its density ρ and its pressure p, and a vector field, namely its world velocity u^μ. In order that ρ be a scalar field, one must define it to be the *proper density*, i.e. the rest mass per unit rest volume. In place of the particle 4-momentum $p^\mu \equiv mu^\mu$, we now have the 4-momentum density ρu^μ.

What we wish to do is to exhibit some tensor which in some way represents the energy content of the fluid, and which, when taken over to curved spacetime, can act as the source of the gravitational field. Since in relativity we lose the distinction between mass and energy, all forms of energy should produce a gravitational field. Moreover, energy is not a scalar, but only the zeroth component of the 4-momentum, so we expect our source tensor to contain the 4-momentum density of the fluid. Rather than try to construct a suitable tensor, let us simply write one down, and then discuss its physical significance. The tensor in question is the *energy-momentum-stress tensor* (or *stress tensor* for short), and for a perfect fluid is defined to be

$$T^{\mu\nu} \equiv (\rho + p/c^2)u^\mu u^\nu - p\eta^{\mu\nu}. \tag{3.1.2}$$

The first thing to note is that $T^{\mu\nu}$ is symmetric, and is made up from ρ, p and u^μ, the scalar and vector fields which characterise the fluid. The pressure of the fluid makes some contribution to its energy content, and so should find a place in the tensor. The next thing to note is that

$$T^{\mu\nu}u_\nu = c^2(\rho + p/c^2)u^\mu - pu^\mu = c^2\rho u^\mu,$$

so $T^{\mu\nu}u_\nu$ is (up to a factor c^2) the 4-momentum density of the fluid. Finally we assert that setting its divergence $T^{\mu\nu}{}_{,\mu}$ equal to zero yields two important equations, namely the continuity equation and the equation of motion. (Since $T^{\mu\nu}$ is symmetric, it has only one divergence.) To prove our assertion would involve us in a lengthy digression into relativistic fluid mechanics, so we will simply derive the two equations, and give supporting arguments for their validity.

Setting $T^{\mu\nu}{}_{,\mu} = 0$ gives

$$(\rho u^\mu)_{,\mu} u^\nu + \rho u^\mu u^\nu{}_{,\mu} + (p/c^2)u^\mu{}_{,\mu} u^\nu + (p/c^2)u^\mu u^\nu{}_{,\mu}$$
$$+ c^{-2}p_{,\mu}u^\mu u^\nu - p_{,\mu}\eta^{\mu\nu} = 0. \tag{3.1.3}$$

Now the world velocity u^ν satisfies $u^\nu u_\nu = c^2$, and differentiation gives

$$u^\nu{}_{,\mu} u_\nu + u^\nu u_{\nu,\mu} = 0, \tag{3.1.4}$$

which implies that $u^\nu{}_{,\mu} u_\nu = 0$ (see Exercise 3.1.3). So contracting equation (3.1.3) with u_ν and dividing by c^2 gives

$$(\rho u^\mu)_{,\mu} + (p/c^2)u^\mu{}_{,\mu} = 0. \tag{3.1.5}$$

Equation (3.1.3) therefore simplifies to

$$(\rho + p/c^2)u^\nu{}_{,\mu} u^\mu = (\eta^{\mu\nu} - c^{-2}u^\mu u^\nu)p_{,\mu}. \tag{3.1.6}$$

Note that in obtaining equation (3.1.5) we contracted the equation $T^{\mu\nu}{}_{,\mu} = 0$ with u_ν, which is equivalent to taking its zeroth component in an instantaneous rest system of the fluid at the point in question.

We now justify our assertion concerning $T^{\mu\nu}{}_{,\mu} = 0$, by arguing that equation (3.1.5) is the equation of continuity of the fluid, while equation (3.1.6) is its equation of motion. We do this by showing that, for slowly moving fluids and small pressures, they reduce to the classical equations. To this end let us put $u^\mu = \gamma v^\mu = \gamma(c, \mathbf{v})$. Then by a slowly moving fluid we mean one for which we may neglect v/c, and so take $\gamma = 1$, and by small pressures we mean that p/c^2 is negligible compared to ρ. Equation (3.1.5) then reduces to

$$(\rho v^\mu)_{,\mu} = 0,$$

or

$$(\rho c)_{,0} + (\rho v^i)_{,i} = 0.$$

In 3-vector notation this is

$$\partial\rho/\partial t + \nabla \cdot (\rho \mathbf{v}) = 0,$$

which is the classical continuity equation [2], the difference between proper density and density disappearing in the classical limit.

As for equation (3.1.6), this reduces to

$$\rho v^\nu{}_{,\mu} v^\mu = (\eta^{\mu\nu} - c^{-2} v^\mu v^\nu) p_{,\mu}, \tag{3.1.7}$$

and in our approximation

$$[\eta^{\mu\nu} - c^{-2} v^\mu v^\nu] = \begin{bmatrix} 0 & 0 & 0 & 0 \\ 0 & -1 & 0 & 0 \\ 0 & 0 & -1 & 0 \\ 0 & 0 & 0 & -1 \end{bmatrix},$$

so the zeroth components of the left-hand and right-hand sides are both zero. Its non-zeroth components reduce to

$$\rho v^i{}_{,\mu} v^\mu = -\delta^{ii} p_{,j},$$

or

$$\rho[\partial v^i/\partial t + v^i{}_{,j} v^j] = -\delta^{ii} p_{,j}.$$

In 3-vector notation this is

$$\rho(\partial/\partial t + \mathbf{v} \cdot \nabla)\mathbf{v} = -\nabla p,$$

which is Euler's classical equation of motion (1755) for a perfect fluid [3].

Returning to the relativistic continuity equation (3.1.5), we see that it contains the pressure as well as the density, but this is not surprising, for we know that in relativity it is energy rather than mass which is conserved, and for a fluid under pressure, the pressure makes a contribution to the energy content. The relativistic equation (3.1.6) may be written in the form

$$(\rho + p/c^2)\, d^2x^\nu/d\tau^2 = (\eta^{\mu\nu} - c^{-2}u^\mu u^\nu)p_{,\mu},$$

for

$$u^\nu{}_{,\mu}\, u^\mu = \left(\frac{\partial}{\partial x^\mu} \frac{dx^\nu}{d\tau} \right) \frac{dx^\mu}{d\tau} = \frac{d^2x^\nu}{d\tau^2}.$$

In this form it looks more like an equation of motion, for it shows that the fluid particles are pushed off geodesics ($d^2x^\nu/d\tau^2 = 0$) by the pressure gradient $p_{,\mu}$.

If we were to accept equation (3.1.5) as the relativistic continuity equation and equation (3.1.6) as the equation of motion of a perfect fluid, then we could reverse our argument, and claim that $T^{\mu\nu}{}_{,\mu} = 0$ by virtue of the continuity equation and the equation of motion. It is, in fact, possible to give more complicated expressions representing the stress tensors of imperfect fluids and charged fluids, and even an electromagnetic field. These tensors are all symmetric, and all have zero divergence by virtue of equations such as continuity equations, equations of motion, or Maxwell's equations (see Problem 3.4).

Let us now take the step to the curved spacetime of general relativity. Our discussion in Section 2.6 gave us a prescription for taking over definitions and tensor equations from flat spacetime. In particular we replace $\eta_{\mu\nu}$ by $g_{\mu\nu}$, and partial by covariant derivatives, so our defining equation for the stress tensor of a perfect fluid becomes

$$T^{\mu\nu} \equiv (\rho + p/c^2)u^\mu u^\nu - pg^{\mu\nu}, \tag{3.1.8}$$

and the vanishing of its divergence is expressed as

$$T^{\mu\nu}{}_{;\mu} = 0. \tag{3.1.9}$$

With suitable definitions of $T^{\mu\nu}$ equation (3.1.9) is valid for all fluids and fields, not just perfect fluids. It is the stress tensor which we take as the source of the gravitational field, and the result (3.1.9) plays an important role in formulating the field equations; but before doing that we must take a closer look at curvature.

Exercises 3.1

1. A particle of 4-momentum p^μ just misses an observer with world velocity U^μ. Show that he assigns an energy $p_\mu U^\mu$ (evaluated at the event of near-collision) to the particle.
2. Check that all the terms on the right-hand side of equation (3.1.2) have the same dimension.
3. Verify that $u^\nu u_\nu = c^2$ implies that $u^\nu{}_{;\mu} u_\nu = 0$.

3.2   The curvature tensor and related tensors

The material of this section is applicable to any N-dimensional Riemannian manifold, so we use suffices a, b, etc., which have the range 1 to N, rather than μ, ν, etc., with the range 0 to 3. Of course, the requirements of general relativity will govern the scope of our results.

We remarked in Section 2.3 that covariant differentiations were not commutative, and we start by taking a closer look at this question. For a covariant vector field λ_a

$$\lambda_{a;b} \equiv \partial_b \lambda_a - \Gamma^d_{ab}\lambda_d,$$

and a further covariant differentiation gives

$$\begin{aligned}
\lambda_{a;bc} &= \partial_c(\lambda_{a;b}) - \Gamma^e_{ac}\lambda_{e;b} - \Gamma^e_{bc}\lambda_{a;e} \\
&= \partial_c\partial_b\lambda_a - (\partial_c\Gamma^d_{ab})\lambda_d - \Gamma^d_{ab}\partial_c\lambda_d \\
&\quad - \Gamma^e_{ac}(\partial_b\lambda_e - \Gamma^d_{eb}\lambda_d) - \Gamma^e_{bc}(\partial_e\lambda_a - \Gamma^d_{ae}\lambda_d).
\end{aligned}$$

Interchanging b and c, and then subtracting gives

$$\lambda_{a;bc} - \lambda_{a;cb} = R^d{}_{abc}\lambda_d, \tag{3.2.1}$$

where

$$R^d_{abc} \equiv \partial_b \Gamma^d_{ac} - \partial_c \Gamma^d_{ab} + \Gamma^e_{ac}\Gamma^d_{eb} - \Gamma^e_{ab}\Gamma^d_{ec}. \qquad (3.2.2)$$

The left-hand side of equation (3.2.1) is a tensor for arbitrary vectors λ_a, so the contraction of R^d_{abc} with λ_d is a tensor, and since R^d_{abc} does not depend on λ_a, the quotient theorem entitles us to conclude that R^d_{abc} is a type $(1, 3)$ tensor. It is called the *curvature tensor* (or *Riemann–Christoffel curvature tensor* or *Riemann tensor*), and equation (3.2.2) indicates that it is defined in terms of the metric tensor and its derivatives.

So the necessary and sufficient condition that covariant differentiations of a type $(0, 1)$ tensor field commute is that $R^a_{bcd} = 0$. This is in fact the necessary and sufficient condition for the commutativity of tensor fields of all types (see Exercise 3.2.1).

In the flat spacetime of special relativity we know that coordinate systems exist in which $g_{\mu\nu} = \eta_{\mu\nu}$, and in these coordinate systems $\Gamma^\mu_{\nu\sigma} = 0$, and hence the curvature tensor is identically zero. However, this does not entitle us to assume that the curvature tensor field of an arbitrary Riemannian manifold is zero, and in Problem 3.1 we give an example of a 2-dimensional manifold with non-vanishing curvature tensor field. We can now give a more formal definition of flatness. A manifold is *flat* if at each point of it $R^a_{bcd} = 0$, otherwise it is *curved*. (We may also speak of flat regions of a manifold.) It may be shown that in any region where $R^a_{bcd} = 0$ it is possible to introduce a coordinate system in which the components g_{ab} are constants, and hence a cartesian coordinate system (one in which $[g_{ab}]$ is a diagonal matrix with ± 1 as its diagonal entries) [4].

On the face of it, R^a_{bcd} has N^4 components. However, it possesses a number of symmetries and its components satisfy a certain identity, and it may be shown that these cut the number down to $N^2(N^2 - 1)/12$ independent components. The identity is given by the relation

$$R^a_{bcd} + R^a_{cdb} + R^a_{dbc} = 0, \qquad (3.2.3)$$

and is known as the *cyclic identity*. Its proof is left as an exercise. The symmetries possessed by the curvature tensor are best expressed in terms of the associated type $(0, 4)$ tensor

$$R_{abcd} \equiv g_{ae}R^e_{bcd}.$$

Making use of equations (2.4.5), (2.4.7) and (2.4.8) gives, after extensive manipulation,

$$R_{abcd} \equiv \tfrac{1}{2}(\partial_d \partial_a g_{bc} - \partial_d \partial_b g_{ac} + \partial_c \partial_b g_{ad} - \partial_c \partial_a g_{bd})$$
$$- g^{ef}(\Gamma_{eac}\Gamma_{fbd} - \Gamma_{ead}\Gamma_{fbc}). \quad (3.2.4)$$

From this form for R_{abcd} it is a simple matter to check the following symmetry properties:

(a)	$R_{abcd} = -R_{bacd},$	(3.2.5a)
(b)	$R_{abcd} = -R_{abdc},$	(3.2.5b)
(c)	$R_{abcd} = R_{cdab}.$	(3.2.5c)

It follows from (a) that

$$R^a{}_{acd} = 0. \quad (3.2.6)$$

The covariant derivatives $R^a{}_{bcd;e}$ also satisfy an identity, namely

$$R^a{}_{bcd;e} + R^a{}_{bde;c} + R^a{}_{bec;d} = 0. \quad (3.2.7)$$

This is known as the *Bianchi identity*, and may easily be proved in the following way. About any point P we can construct a coordinate system with $(\Gamma^a_{bc})_P = 0$. Differentiating equation (3.2.2) and then evaluating at P gives, in this coordinate system,

$$(R^a{}_{bcd;e})_P = (\partial_e \partial_c \Gamma^a_{bd} - \partial_e \partial_d \Gamma^a_{bc})_P.$$

Cyclically permuting c, d and e and adding gives the result at P. But P is arbitrary, so the result holds everywhere.

Equation (3.2.6) states that the contraction $R^a{}_{acd}$ is zero. However, in general the contraction $R^a{}_{bca}$ is non-zero, and this leads to a new tensor, the *Ricci tensor*. It is traditional to use the same kernel letter for the Ricci tensor as for the curvature tensor, so we denote its components by [5]

$$R_{ab} \equiv R^c{}_{abc}. \quad (3.2.8)$$

The Ricci tensor is in fact symmetric, as may be shown by contracting the cyclic identity (see Exercise 3.2.4). Since R_{ab} is symmetric, $R^a{}_b = R_b{}^a$ and we can denote both by R^a_b. A further contraction gives the *curvature scalar*

$$R \equiv g^{ab}R_{ab} = R^a_a, \qquad (3.2.9)$$

and again the same kernel letter is used.

One final tensor, which is of some importance for later work, is the *Einstein tensor* G_{ab} defined by

$$G_{ab} \equiv R_{ab} - \tfrac{1}{2}Rg_{ab}. \qquad (3.2.10)$$

It is clearly symmetric, and this means that it possesses only one divergence $G^{ab}_{;a}$. The reason for the importance of the Einstein tensor is that this divergence is zero. Contracting a with d in the Bianchi identity (3.2.7) gives

$$R_{bc;e} + R^a_{bae;c} + R^a_{bec;a} = 0,$$

or, on using equation (3.2.5b),

$$R_{bc;e} - R_{be;c} + R^a_{bec;a} = 0.$$

If we now raise b and contract with e, we get

$$R^b_{c;b} - R_{;c} + R^{ab}_{bc;a} = 0.$$

But, from equation (3.2.5c),

$$R^{ab}_{bc;a} = R^{ba}_{cb;a} = R^a_{c;a} = R^b_{c;b},$$

so the above reduces to

$$2R^b_{c;b} - R_{;c} = 0,$$

or

$$(R^b_c - \tfrac{1}{2}R\delta^b_c)_{;b} = 0,$$

on dividing by two and using the result of Exercise 2.3.4. We thus have $G^b_{c;b} = 0$, which implies $G^{bc}_{;b} = 0$, as asserted.

One important place in which the curvature makes an appearance is in the equation of geodesic deviation, a concept introduced in Section 2.0. Consider two neighbouring geodesics, γ given by $x^a(u)$ and $\tilde{\gamma}$ given by $\tilde{x}^a(u)$, both affinely parametrised, and let $\xi^a(u)$ be the small "vector" connecting points with the same parameter value, i.e. $\xi^a(u) \equiv \tilde{x}^a(u) - x^a(u)$, (see Fig. 3.1). (We are here making use of the idea put forward in Section 2.2 that small coordinate differences may be regarded as the

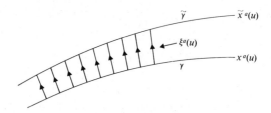

Fig. 3.1   Geodesic deviation.

components of a vector.) To ensure that ξ^a as defined above is small, we must be careful with our parametrisation. Having parametrised γ, we can use the affine freedom available $(u \rightarrow Au + B)$ to parametrise $\tilde{\gamma}$ so that (over some range of parameter values) ξ^a is indeed small. If neither geodesic is null, then arc-length s may be used, and care need only be taken in fixing the zero value of s on, say, $\tilde{\gamma}$.

Since γ and $\tilde{\gamma}$ are geodesics we have

$$\frac{\mathrm{d}^2 \tilde{x}^a}{\mathrm{d}u^2} + \tilde{\Gamma}^a_{bc} \frac{\mathrm{d}\tilde{x}^b}{\mathrm{d}u} \frac{\mathrm{d}\tilde{x}^c}{\mathrm{d}u} = 0, \tag{3.2.11}$$

and

$$\frac{\mathrm{d}^2 x^a}{\mathrm{d}u^2} + \Gamma^a_{bc} \frac{\mathrm{d}x^b}{\mathrm{d}u} \frac{\mathrm{d}x^c}{\mathrm{d}u} = 0, \tag{3.2.12}$$

where the tilde on the connection coefficient in equation (3.2.11) indicates that it is evaluated at the point with coordinates $\tilde{x}^a(u)$, whereas that in equation (3.2.12) is evaluated at the point with coordinates $x^a(u)$. But to first order in ξ^a,

$$\tilde{\Gamma}^a_{bc} = \Gamma^a_{bc} + \Gamma^a_{bc,d}\xi^d,$$

and subtracting equation (3.2.12) from (3.2.11) gives

$$\ddot{\xi}^a + \Gamma^a_{bc,d}\dot{x}^b\dot{x}^c\xi^d + \Gamma^a_{bc}\dot{x}^b\dot{\xi}^c + \Gamma^a_{bc}\dot{\xi}^b\dot{x}^c = 0,$$

where dots denote derivatives with respect to u, and only first-order terms have been retained. This may be written as

$$\mathrm{d}(\dot{\xi}^a + \Gamma^a_{bc}\xi^b\dot{x}^c)/\mathrm{d}u - \Gamma^a_{bc,d}\xi^b\dot{x}^c\dot{x}^d - \Gamma^a_{bc}\xi^b\ddot{x}^c$$
$$+ \Gamma^a_{bc,d}\dot{x}^b\dot{x}^c\xi^d + \Gamma^a_{bc}\dot{x}^b\dot{\xi}^c = 0.$$

Substitution for $\ddot{x}^c$ from equation (3.2.12) and some rearrangement gives

$$d(\dot{\xi}^a + \Gamma^a_{bc}\xi^b\dot{x}^c)/du + \Gamma^a_{de}(\dot{\xi}^d + \Gamma^d_{bc}\xi^b\dot{x}^c)\dot{x}^e$$
$$- \Gamma^a_{de}\Gamma^d_{bc}\xi^b\dot{x}^c\dot{x}^e - \Gamma^a_{bc,d}\xi^b\dot{x}^c\dot{x}^d$$
$$+ \Gamma^a_{bc}\xi^b\Gamma^c_{de}\dot{x}^d\dot{x}^e + \Gamma^a_{bc,d}\dot{x}^b\dot{x}^c\xi^d = 0.$$

On relabelling dummy suffices, this rather complicated expression reduces to

$$D^2\xi^a/du^2 + (\Gamma^a_{cd,b} - \Gamma^a_{bc,d} + \Gamma^a_{be}\Gamma^e_{dc} - \Gamma^a_{ec}\Gamma^e_{bd})\xi^b\dot{x}^c\dot{x}^d = 0,$$

or

$$D^2\xi^a/du^2 + R^a{}_{cbd}\xi^b\dot{x}^c\dot{x}^d = 0. \tag{3.2.13}$$

This is the *equation of geodesic deviation*.

In a flat manifold $R^a{}_{bcd} = 0$, and in cartesian coordinates $D/du = d/du$, so equation (3.2.13) then reduces to $d^2\xi^a/du^2 = 0$, which implies that $\xi^a(u) = A^a u + B^a$, where the A^a and B^a are constants. So in a flat manifold the separation vector increases linearly with u (and therefore with s if γ is non-null). However, in a curved manifold $R^a{}_{bcd} \neq 0$, and we do not have this linear relationship. These observations should be compared with the remarks made in Section 2.0.

Exercises 3.2

1. (a) Show that for a contravariant vector field λ^a,

$$\lambda^a{}_{;bc} - \lambda^a{}_{;cb} = -R^a{}_{dbc}\lambda^d.$$

(b) Show that for a type $(2, 0)$ tensor field τ^{ab},

$$\tau^{ab}{}_{;cd} - \tau^{ab}{}_{;dc} = -R^a{}_{ecd}\tau^{eb} - R^b{}_{ecd}\tau^{ae}.$$

(Without loss of generality take $\tau^{ab} = \lambda^a\mu^b$.)

(c) Guess the corresponding expression for a type $(2, 1)$ tensor field τ^{ab}_c.

2. Prove the cyclic identity (3.2.3).

3. Verify the form of R_{abcd} given by equation (3.2.4).

4. By contracting the cyclic identity (3.2.3) prove that the Ricci tensor is symmetric.

5. Check the derivation of equation (3.2.13).

3.3   Einstein's field equations

The field equations of general relativity are variously referred to as *Einstein's equation* or *Einstein's field equations*, and they were obtained by him at the end of 1915, after what he referred to as a period of unremitting labour. During the approximate period 1909–13 Einstein and his friend from undergraduate days, Marcel Grossmann, had realised that the metric tensor $g_{\mu\nu}$ describing the geometry of spacetime seemed to depend on the amount of gravitating matter in the region in question (and so adopted the kernel letter g for *gravity*) [6].

The metric tensor contains two separate pieces of information:

(i) the relatively unimportant information concerning the specific coordinate system used (e.g. spherical polars, cartesians, etc.);
(ii) the important information regarding the existence of any gravitational potentials.

In Section 2.8, we saw that in a nearly cartesian coordinate system g_{00} was essentially the Newtonian potential. In a more general coordinate system, this Newtonian potential would be dispersed throughout the $g_{\mu\nu}$, so there is a sense in which all the components $g_{\mu\nu}$ can be regarded as gravitational potentials.

We have seen in Section 3.1 that the matter content of spacetime is concisely summarised in the stress tensor $T^{\mu\nu}$, so if matter causes the geometry, it might be tempting to put

$$g^{\mu\nu} = \kappa T^{\mu\nu}, \tag{3.3.1}$$

where κ is some coupling constant. This looks plausible, because both $g^{\mu\nu}$ and $T^{\mu\nu}$ are symmetric, and $g^{\mu\nu}{}_{;\mu} = 0$ (see Exercise 2.4.2) in agreement with $T^{\mu\nu}{}_{;\mu} = 0$. However, equation (3.3.1) does not reduce to Poisson's equation, $\nabla^2 V = 4\pi G\rho$, in the Newtonian limit. Since the $g_{\mu\nu}$ are the gravitational potentials, it is clear that what is needed in place of $g^{\mu\nu}$ in equation (3.3.1) is a symmetric tensor involving the second derivatives of $g_{\mu\nu}$.

During the period 1914–15 Einstein made many attempts to find the exact form of the suspected relationship between the metric tensor and matter, and in 1915 (by which time he had moved to Berlin, leaving Grossmann in Zurich) he published his belief in the equation

$$R^{\mu\nu} = \kappa T^{\mu\nu}, \tag{3.3.2}$$

where $R^{\mu\nu}$ is the contravariant Ricci tensor. Again this looks plausible, since $R^{\mu\nu}$ is symmetric and contains second derivatives of $g_{\mu\nu}$. However, $R^{\mu\nu}$ does not satisfy $R^{\mu\nu}{}_{;\mu} = 0$, and later in the same year he modified the equation to

$$R^{\mu\nu} - \tfrac{1}{2}Rg^{\mu\nu} = \kappa T^{\mu\nu}. \tag{3.3.3}$$

The left-hand side of this equation is the Einstein tensor $G^{\mu\nu}$, and we know from the previous section that $G^{\mu\nu}{}_{;\mu} = 0$, so equation (3.3.3) looks satisfactory in all respects. We shall see in the next section that it gives Poisson's equation as an approximation, and that this approximation allows us to give the coupling constant κ the value $-8\pi G/c^4$. (Note that we now have ten field equations replacing the single field equation of the Newtonian theory.) An alternative form for the field equations (3.3.3) is

$$R^{\mu\nu} = \kappa(T^{\mu\nu} - \tfrac{1}{2}Tg^{\mu\nu}), \tag{3.3.4}$$

where $T \equiv T^{\mu}_{\mu}$ (see Exercise 3.3.1).

Recall that $T^{\mu\nu}$ contains all forms of energy and momentum. For example, if there is electromagnetic radiation present, then this must be included in $T^{\mu\nu}$. A region of spacetime in which $T^{\mu\nu} = 0$ is called *empty*, and such region is therefore not only devoid of matter, but of radiative energy and momentum also. It can be seen from equation (3.3.4) that the empty-spacetime field equations are

$$R^{\mu\nu} = 0. \tag{3.3.5}$$

Further support for the correctness of the field equations is given by comparing the equation of geodesic deviation with its Newtonian counterpart. With proper time τ as affine parameter, the equation (3.2.13) of geodesic deviation takes the form

$$D^2\xi^\mu/d\tau^2 + R^\mu{}_{\sigma\nu\rho}\xi^\nu\dot{x}^\sigma\dot{x}^\rho = 0, \tag{3.3.6}$$

where $\xi^\mu(\tau)$ is the small vector connecting corresponding points on neighbouring geodesics. For comparison with its Newtonian counterpart, let us write this as

$$D^2\xi^\mu/d\tau^2 = -K^\mu_\nu\xi^\nu, \tag{3.3.7}$$

where

$$K^\mu_\nu \equiv R^\mu{}_{\sigma\nu\rho}\dot{x}^\sigma\dot{x}^\rho = -R^\mu{}_{\sigma\rho\nu}\dot{x}^\sigma\dot{x}^\rho. \tag{3.3.8}$$

The corresponding situation in Newtonian gravitation theory is two particles moving under gravity on neighbouring paths given by $\tilde{x}^i(t)$ and $x^i(t)$. Their equations of motion are

$$\mathrm{d}^2\tilde{x}^i/\mathrm{d}t^2 = -\delta^{ik}\, \tilde{\partial}_k V,$$

and

$$\mathrm{d}^2 x^i/\mathrm{d}t^2 = -\delta^{ik}\, \partial_k V,$$

where $\tilde{\partial}_k$ in the first equation indicates that the gradient of the gravitational potential V is evaluated at $\tilde{x}^i(t)$. If we subtract and put $\xi^i(t) \equiv \tilde{x}^i(t) - x^i(t)$, and make use of the fact that, for small ξ^j,

$$\tilde{\partial}_k V = \partial_k V + (\partial_j \partial_k V)\xi^j,$$

then there results

$$\mathrm{d}^2\xi^i/\mathrm{d}t^2 = -\delta^{ik}(\partial_j \partial_k V)\xi^j,$$

or

$$\mathrm{d}^2\xi^i/\mathrm{d}t^2 = -K^i_j \xi^j, \tag{3.3.9}$$

where

$$K^i_j \equiv \delta^{ik}\partial_j \partial_k V. \tag{3.3.10}$$

Equation (3.3.9) is the Newtonian counterpart of equation (3.3.7) and brings out the correspondence:

$$K^\mu_\nu \equiv -R^\mu{}_{\sigma\rho\nu}\dot{x}^\sigma\dot{x}^\rho \leftrightarrow K^i_j \equiv \delta^{ik}\, \partial_j \partial_k V.$$

Now the empty-space field equation of Newtonian gravitation is $\nabla^2 V = 0$, or equivalently $K^i_i = 0$. This suggests that in empty spacetime we should have $K^\mu_\mu = 0$, i.e. $R_{\sigma\rho}\dot{x}^\sigma\dot{x}^\rho = 0$. Since this should hold for arbitrary tangent vectors $\dot{x}^\mu$ to geodesics we conclude (because $R_{\mu\nu}$ is symmetric) that $K^\mu_\mu = 0$ is equivalent to $R_{\sigma\rho} = 0$. In this way comparison between the equation of geodesic deviation and its Newtonian counterpart lends support to equation (3.3.5) as the field equations of empty spacetime. Support for the non-empty spacetime field equations (3.3.3) or (3.3.4) is given in the next section.

Exercise 3.3

1. By contracting the mixed form

$$R^\mu_\nu - \tfrac{1}{2}R\delta^\mu_\nu = \kappa T^\mu_\nu$$

of equation (3.3.3) show that $R = -\kappa T$, where $T \equiv T^\mu_\mu$, and hence verify equation (3.3.4).

3.4 Einstein's equation compared with Poisson's equation

Poisson's equation may be recovered from Einstein's equation by considering its 00-component in the 'weak-field' approximation. We use the covariant version of equation (3.3.4), and so are interested in

$$R_{00} = \kappa(T_{00} - \tfrac{1}{2}Tg_{00}). \tag{3.4.1}$$

As in Section 2.8, we use a nearly cartesian coordinate system in which $g_{\mu\nu} = \eta_{\mu\nu} + h_{\mu\nu}$, where products of the $h_{\mu\nu}$ may be neglected; we also assume that the quasi-static condition of that section holds.

Let us assume that our weak gravitational field arises from a perfect fluid whose particles have (in our coordinate system) speeds v which are small when compared with c, so we take $\gamma = (1 - v^2/c^2)^{-1/2}$ to be one. For most classical distributions (e.g. water, the Sun, or a gas at high pressure) $p/c^2 \ll \rho$, so we take for the stress tensor

$$T_{\mu\nu} = \rho u_\mu u_\nu.$$

This gives $T = \rho c^2$, and equation (3.4.1) becomes

$$R_{00} = \kappa\rho(u_0 u_0 - \tfrac{1}{2}c^2 g_{00}).$$

But $u_0 \simeq c$ and $g_{00} \simeq 1$, so we have

$$R_{00} \simeq \tfrac{1}{2}\kappa\rho c^2, \tag{3.4.2}$$

where, from equation (3.2.2),

$$R_{00} = \partial_0 \Gamma^\mu_{0\mu} - \partial_\mu \Gamma^\mu_{00} + \Gamma^\nu_{0\mu}\Gamma^\mu_{\nu 0} - \Gamma^\nu_{00}\Gamma^\mu_{\nu\mu}. \tag{3.4.3}$$

In our nearly cartesian coordinate system the $\Gamma^\mu_{\nu\sigma}$ are small, so we can neglect the last two terms in equation (3.4.3) [7], and on using the quasi-static condition we have

$$R_{00} \simeq -\partial_i \Gamma^i_{00}.$$

But from Section 2.8, we have that in this approximation

$$\Gamma^i_{00} = \tfrac{1}{2}\delta^{ij}\,\partial_j h_{00},$$

so equation (3.4.2) reduces to

$$-\tfrac{1}{2}\delta^{ij}\,\partial_i\,\partial_j h_{00} \simeq \tfrac{1}{2}\kappa\rho c^2.$$

But $\delta^{ij}\,\partial_i\,\partial_j = \nabla^2$, and from equation (2.8.10) $h_{00} = 2V/c^2$, where V is the gravitational potential, so there results

$$\nabla^2 V \simeq -\tfrac{1}{2}\kappa\rho c^4, \tag{3.4.4}$$

which corresponds satisfactorily with Poisson's equation, provided we identify the coupling constant κ in Einstein's equation as $-8\pi G/c^4$. Equation (3.4.4) then becomes

$$\nabla^2 V \simeq 4\pi G\rho.$$

3.5   The Schwarzschild solution

Is it possible to solve the field equations and thus discover $g_{\mu\nu}$? If one examines how $g_{\mu\nu}$ enters $R^{\mu\nu}$ and $G^{\mu\nu}$, one readily appreciates the high degree of non-linearity possessed by the equations, so any solution will not be easy to obtain. The problem becomes easier if one looks for special solutions, for example those representing spacetimes possessing symmetries, and the first exact solution, obtained by K. Schwarzschild in 1916, is one of this type.

What Schwarzschild sought was the metric tensor field representing the static spherically symmetric gravitational field in the empty spacetime surrounding some massive spherical object like a star. His guiding assumptions were [8]:

(a) that the field was static,
(b) that the field was spherically symmetric,
(c) that the spacetime was empty,
(d) that the spacetime was asymptotically flat.

He also assumed that spacetime could be coordinatised by coordinates (t, r, θ, ϕ), where t was a timelike coordinate [9], θ and ϕ were polar angles picking out radial directions in the usual manner, and r was some radial coordinate, and he then postulated

$$c^2\,d\tau^2 = A(r)\,dt^2 - B(r)\,dr^2 - r^2\,d\theta^2 - r^2\sin^2\theta\,d\phi^2 \tag{3.5.1}$$

as a form for the line element, where $A(r)$ and $B(r)$ were some unknown functions of r to be obtained by solving the field equations.

We shall not give the reasons why Schwarzchild arrived at the form (3.5.1) for the line element, but we can make some observations concerning it. The fact that none of the $g_{\mu\nu}$ depends on t expresses his assumption (a), and the fact that the surfaces given by r, t constant have line elements

$$\mathrm{d}s^2 = r^2(\mathrm{d}\theta^2 + \sin^2\theta\,\mathrm{d}\phi^2), \qquad (3.5.2)$$

and so have the geometry of spheres (see Problem 2.1), expresses his assumption (b). Assumption (c) means that $A(r)$ and $B(r)$ are to be found using the empty-spacetime field equations $R_{\mu\nu} = 0$, while assumption (d) gives boundary conditions on A and B, namely

$$A(r) \to c^2 \quad \text{and} \quad B(r) \to 1 \quad \text{as} \quad r \to \infty \qquad (3.5.3)$$

(see Problem 2.1). Note that because $B(r)$ is not necessarily 1, we cannot assume that r is radial distance. In fact the line element (3.5.2) shows that a surface given by r, t constant has surface area $4\pi r^2$, and at the moment this is the only meaning we can give to r; further discussion on its meaning is given in the next chapter.

Let us now retrace Schwarzschild's solution of the field equations. The idea is to use $g_{\mu\nu}$ obtained from the line element (3.5.1) as a trial solution for the empty-spacetime field equations. As with all trial solutions, the main justification for it is that it works. From equation (3.2.2) we have

$$R_{\mu\nu} \equiv \partial_\nu \Gamma^\sigma_{\mu\sigma} - \partial_\sigma \Gamma^\sigma_{\mu\nu} + \Gamma^\rho_{\mu\sigma} \Gamma^\sigma_{\rho\nu} - \Gamma^\rho_{\mu\nu} \Gamma^\sigma_{\rho\sigma},$$

and from Problem 2.7 we have

$$\Gamma^0_{01} = A'/2A, \qquad \Gamma^1_{00} = A'/2B, \qquad \Gamma^1_{11} = B'/2B,$$

$$\Gamma^1_{22} = -r/B, \qquad \Gamma^1_{33} = -(r\sin^2\theta)/B, \qquad \Gamma^2_{12} = 1/r,$$

$$\Gamma^2_{33} = -\sin\theta\cos\theta, \qquad \Gamma^3_{13} = 1/r, \qquad \Gamma^3_{23} = \cot\theta,$$

all other connection coefficients being zero. Here we have labelled the coordinates according to $x^0 \equiv t$, $x^1 \equiv r$, $x^2 \equiv \theta$, $x^3 \equiv \phi$, and a prime denotes differentiation with respect to r. Tedious substitution then shows that $R_{\mu\nu} = 0$ gives (see Exercise 3.5.1):

$$R_{00} = -\frac{A''}{2B} + \frac{A'}{4B}\left(\frac{A'}{A} + \frac{B'}{B}\right) - \frac{A'}{rB} = 0, \tag{3.5.4a}$$

$$R_{11} = \frac{A''}{2A} - \frac{A'}{4A}\left(\frac{A'}{A} + \frac{B'}{B}\right) - \frac{B'}{rB} = 0, \tag{3.5.4b}$$

$$R_{22} = \frac{1}{B} - 1 + \frac{r}{2B}\left(\frac{A'}{A} - \frac{B'}{B}\right) = 0, \tag{3.5.4c}$$

$$R_{33} = R_{22}\sin^2\theta = 0. \tag{3.5.4d}$$

Fortunately, $R_{\mu\nu} = 0$ identically for $\mu \neq \nu$.

Of the four equations (3.5.4), only the first three are useful. Adding B/A times (3.5.4a) to (3.5.4b) gives

$$\frac{A'}{A} + \frac{B'}{B} = 0, \quad \text{or} \quad A'B + AB' = 0,$$

which implies that $AB = \text{constant}$. We can identify this constant as c^2 from the boundary condition (3.5.3), so

$$AB = c^2, \quad \text{and} \quad B = c^2/A.$$

Substitution in equation (3.5.4c) then gives

$$A + rA' = c^2 \quad \text{or} \quad \mathrm{d}(rA)/\mathrm{d}r = c^2.$$

Integrating, we have

$$rA = c^2(r + k),$$

where k is constant, so

$$A(r) = c^2(1 + k/r) \quad \text{and} \quad B(r) = (1 + k/r)^{-1}.$$

In solving for A and B we have used only the sum of equations (3.5.4a) and (3.5.4b), but not the equations separately. However, it is a simple matter to check that, with these forms for A and B, the equations are satisfied separately. Thus we have solved the field equations, and obtained Schwarzschild's solution in the form

$$c^2\,\mathrm{d}\tau^2 = c^2(1 + k/r)\,\mathrm{d}t^2 - (1 + k/r)^{-1}\,\mathrm{d}r^2 - r^2\,\mathrm{d}\theta^2 - r^2\sin^2\theta\,\mathrm{d}\phi^2,$$

where k is a constant, which we now proceed to identify. It clearly must in some way represent the mass of the object producing the gravitational field.

In the region of spacetime where k/r is small (i.e. in the asymptotic region) the line element differs but little from that of flat spacetime in spherical polar coordinates, so here r is approximately radial distance, the approximation getting better as $r \to \infty$. Moreover, if we put

$$x^0 \equiv ct, \qquad x^1 \equiv r \sin \theta \cos \phi, \qquad x^2 \equiv r \sin \theta \sin \phi, \qquad x^3 \equiv r \cos \theta,$$
$$(3.5.5)$$

we obtain a metric tensor of the form $g_{\mu\nu} = \eta_{\mu\nu} + h_{\mu\nu}$ (see Exercise 3.5.2), where in the asymptotic region the $h_{\mu\nu}$ are small, and $h_{00} = k/r$. But in this region, where r is approximately radial distance, the corresponding Newtonian potential is $V = -MG/r$, where M is the mass of the body producing the field, and G is the gravitational constant. Since $h_{00} \equiv 2V/c^2$, we conclude that $k = -2MG/c^2 r$, and Schwarzschild's solution for the empty spacetime outside a spherical body of mass M is

$$c^2 \, d\tau^2 = c^2 (1 - 2MG/c^2 r) \, dt^2 - (1 - 2MG/c^2 r)^{-1} \, dr^2$$
$$- r^2 \, d\theta - r^2 \sin^2 \theta \, d\phi^2. \quad (3.5.6)$$

This solution is the basis of our discussions in the next chapter.

Exercises 3.5

1. Check the expressions given for $R_{\mu\nu}$ in equations (3.5.4), and that $R_{\mu\nu} = 0$ for $\mu \neq \nu$.
2. If in Schwarzschild's solution we introduce coordinates x^μ defined by equations (3.5.5), what form does $g_{\mu\nu}$ take?

Problems 3

1. Show that in a 2-dimensional Riemannian manifold all components of R_{abcd} are either zero or $\pm R_{1212}$.
   In terms of the usual polar angles (see Problem 2.1) the metric tensor field of a sphere of radius a is given by

$$[g_{ab}] = \begin{bmatrix} a^2 & 0 \\ 0 & a^2 \sin^2 \theta \end{bmatrix}.$$

Show that $R_{1212} = a^2 \sin^2 \theta$, and hence deduce that

$$[R_{ab}] = \begin{bmatrix} -1 & 0 \\ 0 & -\sin^2 \theta \end{bmatrix}$$

and $R = -2/a^2$.

2. In a certain N-dimensional Riemannian manifold the covariant curvature tensor may be expressed as

$$R_{abcd} = g_{ac}S_{bd} + g_{bd}S_{ac} - g_{ad}S_{bc} - g_{bc}S_{ad},$$

where S_{ab} is a type $(0, 2)$ tensor. Show that, provided $N > 2$, $S_{ab} = S_{ba}$, and that, provided $N > 3$, $S_{ab;c} = S_{ac;b}$.

3. *Dust* is a fluid without internal stress or pressure, so its stress tensor is $T^{\mu\nu} \equiv \rho u^\mu u^\nu$. Show that $T^{\mu\nu}{}_{;\mu} = 0$ implies that the dust particles follow geodesics.

4. Let

$$E^{\mu\nu} \equiv -\mu_0^{-1}[F^{\rho\mu}F_\rho{}^\nu - \tfrac{1}{4}g^{\mu\nu}(F_{\rho\sigma}F^{\rho\sigma})],$$

where $F^{\mu\nu}$ is the free-space electromagnetic field tensor. Show that, by virtue of Maxwell's equations

$$E^{\mu\nu}{}_{;\mu} = F^\nu{}_\rho j^\rho,$$

where j^ρ is the 4-current density.

If the stress tensor for a charged unstressed fluid in free space is defined to be

$$T^{\mu\nu} \equiv \mu u^\mu u^\nu + E^{\mu\nu},$$

where μ is its proper density (rather than ρ, to avoid confusion with charge density) and u^μ its world velocity, show that $T^{\mu\nu}{}_{;\mu} = 0$ by virtue of the continuity equation (for matter) and the equation of motion of the fluid.
(See Section A.8 for the relevant definitions and equations, but adapt them to curved spacetime.)

5. In Section 3.5 we remarked that the field equations of general relativity were non-linear. Explain why this is not surprising. Does the principle of superposition hold for solutions of the field equations?
If not, why not?

6. Show that the Schwarzschild line element (3.5.6) may be put

into the *isotropic form*

$$c^2 \, d\tau^2 = c^2 \left(1 - \frac{GM}{2\rho c^2}\right)^2 \left(1 + \frac{GM}{2\rho c^2}\right)^{-2} dt^2$$

$$- \left(1 + \frac{GM}{2\rho c^2}\right)^4 (d\rho^2 + \rho^2 \, d\theta^2 + \rho^2 \sin^2 \theta \, d\phi^2),$$

where the new coordinate ρ is defined by

$$r \equiv \rho \left(1 + \frac{GM}{2\rho c^2}\right)^2.$$

Notes

1. We use m rather than the more usual notation m_0 for rest mass. Similarly we use ρ rather than ρ_0 or ρ_{00} for the proper rest-mass density.
2. See, for example, Landau and Lifshitz, 1959, §1.
3. See, for example, Landau and Lifshitz, 1959, §2.
4. See, for example, Møller, 1972, Appendix 5.
5. There is wide disagreement over the sign of the curvature tensor, many authors giving it the opposite sign to ours. There is less disagreement over the sign of the Ricci tensor, agreement being effected by defining it according to $R_{ab} = R^c{}_{acb}$ in the case of opposite sign.
6. The story of Einstein's quest for the field equations is told in Hoffmann, 1972, Ch. 8.
7. This is equivalent to assuming that the derivatives of the $h_{\mu\nu}$ are also small (see Note 10 of Ch. 2), and it is this assumption which makes the field "weak".
8. Birkhoff's theorem (see, for example, Misner, Thorne and Wheeler, 1973, §32.2) states that (b) and (c) imply (a), so condition (a) is, in fact, redundant.
9. A coordinate is *timelike* if the tangent vector to its coordinate curve is timelike. *Null* and *spacelike coordinates* are correspondingly defined.

Physics in the vicinity of a massive object

4.0 Introduction

In Chapter 3 we obtained the static spherically symmetric solution of Schwarzschild, and identified it as representing the gravitational field surrounding a spherically symmetric body of mass M situated in an otherwise empty spacetime. This solution is asymptotically flat, and in no way incorporates the gravitational effects of distant matter in the universe. Nevertheless, it seems reasonable to adopt it as a model for the gravitational field in the vicinity of a spherical massive object such as a star, where the star's mass is the principal contributor to the gravitational field.

Suppose, somehow, that we are watching the trajectories of laser beams and particles in the vicinity of a star, all of these trajectories being displayed on a large television screen with the star a rather small dot in the middle. If there is a "mass-control" knob which controls only the mass M of the star, we are really asking in this chapter what happens when we turn the knob so as to increase M. With M turned right down to zero, the Schwarzschild line element reduces to that of flat spacetime in spherical polar coordinates. The coordinates t and r then have simple physical meanings: t is the time as measured by clocks which are stationary in the reference system employed, and r is the radial distance from the origin. Turning M up introduces curvature, so that spacetime is no longer flat, and there is no reason to assume that the coordinates have the simple physical meanings they had in flat spacetime. The relationship between coordinates and physically observable quantities is investigated in Section 4.1.

The Schwarzschild solution is the basis for four of the tests of general relativity listed in the Introduction, namely perihelion advance, the bending of light, time delay in radar sounding, and the geodesic effect [1]. The third of these may be discussed without a detailed knowledge of the geodesics, and this we do in Section

4.2. The question of perihelion advance and the bending of light does require some knowledge of the geodesics, and these matters are discussed in Sections 4.4 to 4.6.

Spectral shift is more a test of the principle of equivalence than of general relativity, but inasmuch as the latter is based on the former, it does yield a test of the general theory, and it is appropriate to discuss it in the context of the Schwarzschild solution. This we do in Section 4.3.

The fifth test mentioned in the Introduction has (at the time of writing) yet to be carried out. Satellite experiments have been proposed, and the test should be made in the not-too-distant future. We consider the theory behind this test in Section 4.7.

Before embarking on our detailed discussion, let us say something about the ranges of the coordinates appearing in the Schwarzschild solution. Inasmuch as the metric tensor components $g_{\mu\nu}$ do not depend on t, the solution is static, and we can take $-\infty < t < \infty$. The coordinates θ and ϕ pick out radial directions in the usual manner of polar angles. In accordance with our definition of a chart, a coordinate function is a one-to-one map onto an open set, and consequently we should limit θ and ϕ to the ranges $0 < \theta < \pi$, $0 < \phi < 2\pi$. However, no trouble will be caused if we let θ take the values 0 and π, nor if we let ϕ extend beyond the quoted range, provided we identify the event with coordinates (t, r, θ, ϕ_1) with that with coordinates (t, r, θ, ϕ_2) whenever ϕ_1 and ϕ_2 differ by a multiple of 2π. The radial coordinate r can decrease from infinity until it reaches either the value r_B corresponding to the boundary of the object, or the value $2GM/c^2$, if r_B is not reached first. The reason for the first lower bound is that the solution we have obtained is the *exterior solution*, valid only where the empty-spacetime field equations hold. The reason for the second one is that as r tends to $2GM/c^2$, the component g_{11} of the metric tensor tends to infinity (see the line element (3.5.6)). So the range of r is $r_B < r < \infty$ or $2GM/c^2 < r < \infty$, as appropriate. Should r decrease to $2GM/c^2$ without r_B being reached, then the object is a *black hole*, and we discuss this situation in Section 4.8. In order to be able to step over the threshold at $r = 2GM/c^2$, we must introduce a coordinate system different from that used to derive Schwarzschild's solution.

The chapter finishes with a brief consideration of some other coordinate systems used in connection with the Schwarzschild solution.

4.1   Length and time

The Schwarzschild spacetime has line element

$$c^2 \, d\tau^2 = (1 - 2m/r)c^2 \, dt^2 - (1 - 2m/r)^{-1} \, dr^2 - r^2 \, d\theta^2 - r^2 \sin^2 \theta \, d\phi^2,$$
(4.1.1)

where for convenience we have put $m \equiv GM/c^2$. If we take a slice given by $t = \text{constant}$ we obtain a 3-dimensional manifold with line element

$$ds^2 = (1 - 2m/r)^{-1} \, dr^2 + r^2 \, d\theta^2 + r^2 \sin^2 \theta \, d\phi^2, \qquad (4.1.2)$$

obtained by putting $dt = 0$ in equation (4.1.1). Putting

$$ds^2 = \tilde{g}_{ij} \, dx^i \, dx^j \quad (i, j = 1, 2, 3, \, x^1 \equiv r, \, x^2 \equiv \theta, \, x^3 \equiv \phi),$$

so that $\tilde{g}_{ij} \equiv -g_{ij}$, we see that $\tilde{g}_{ij}$ is a positive-definite metric tensor field on this 3-dimensional manifold, so the slice is a *space* rather than a spacetime. Moreover, no $\tilde{g}_{ij}$ depends on t, so the spaces given by $t = \text{constant}$ have an enduring permanence which allows us to refer to events with the same r, θ, ϕ-coordinates, but different t-coordinates, as occurring at the *same point* in space. We may also speak of *fixed points* in space. This splitting of spacetime into space and time is possible in any static spacetime, but is not a feature of spacetimes in general, and it should be borne in mind that because of this there are fewer problems of definition and identification in static spacetimes than in non-static ones.

   If we turn M (or m) down to zero, then the line element (4.1.1) becomes that of flat spacetime in spherical polar coordinates, while the line element (4.1.2) becomes that of euclidean space in spherical polar coordinates (see Problem 2.1). Turning M up introduces distortion into both spacetime and space, so that neither is flat. This distortion is effectively measured by the dimensionless quantity $2m/r$ occurring in the two line elements, and is greatest when r is least, i.e. when $r = r_B$, the value of r at the boundary of the object, assuming it is not a black hole. For the Earth, $2m/r_B$ is about 10^{-9}, for the Sun it is about 10^{-6}, but for a proton it is as low as 10^{-36}. For white dwarfs, however, it is not negligible, and for typical neutron stars it can be as much as 10–15 per cent.

   In the flat spacetime given by $m = 0$, the coordinate r is simply the distance from the origin, but if we turn m up, things are not so simple, for r then has a positive lower bound (see previous section) and our origin has disappeared. What then does r represent? If we take the sphere in space given by $r = $ constant, its line element is

$$dL^2 = r^2(d\theta^2 + \sin^2 \theta \, d\phi^2), \qquad (4.1.3)$$

obtained by putting $dr = 0$ in the line element (4.1.2). Comparison with the line element of Problem 2.1(b) shows that this sphere has the 2-dimensional geometry of a sphere of radius r embedded in euclidean space, and just as in the flat space, infinitesimal tangential distances are given by

$$dL \equiv r(d\theta^2 + \sin^2 \theta \, d\phi^2)^{1/2}. \qquad (4.1.4)$$

But what about radial distances given by θ and ϕ constant? The line element shows that for these the infinitesimal radial distance is

$$dR \equiv (1 - 2m/r)^{-1/2} \, dr, \qquad (4.1.5)$$

so $dR > dr$ and *r no longer measures radial distance*. The apparent incompatibility of the distances (4.1.4) and (4.1.5) is explained by the curvature of space. In Fig. 4.1, the flat disc S_0 represents a portion of the flat space (m turned down to zero), while the curved surface S_m represents a portion of the curved space (m turned up). The circles C_1 and C_3 represent spheres having the geometry of a sphere of radius r in euclidean space, while C_2 and C_4 represent neighbouring spheres having the geometry of a sphere of radius $r + dr$ in euclidean space. However, it is only in the flat space represented by S_0 that the measured radial distance between the spheres is dr. In the curved space represented by S_m the measured distance is dR given by equation (4.1.5), and this exceeds dr. If we were to measure the circumference of a great circle of the sphere $r = $ constant using small measuring rods, then the same number of rods would be needed in flat space as in the curved space. On the other hand, if we were to measure the radial distance between points with radial coordinates r_1 and r_2, then more rods would be needed in the curved space than in the flat space (see Examples 1 and 2, below, and Fig. 4.2).

Let us now turn our attention to time. One of the basic assumptions taken over from special relativity is that clocks record

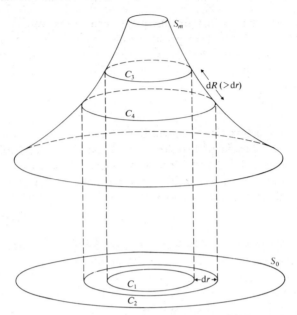

Fig. 4.1   Radial distance in the Schwarzschild geometry.

proper time intervals along their world lines. Infinitesimal proper time intervals are given by the line element (4.1.1), and for a clock at a fixed point in space (r, θ, ϕ constant) this gives

$$d\tau = (1 - 2m/r)^{1/2} \, dt. \qquad (4.1.6)$$

So in flat spacetime (m turned down to zero) $d\tau = dt$, and such a clock records the coordinate time t. However, in the curved spacetime (m turned up) $d\tau < dt$, and fixed clocks do not record coordinate time.

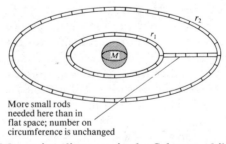

More small rods
needed here than in
flat space; number on
circumference is unchanged

Fig. 4.2   Measuring distances in the Schwarzschild geometry.

Table 4.1.   Comparison of length and time

Special relativity v constant	Schwarzschild r variable
$dl = dl_0(1 - v^2/c^2)^{1/2}$ $dt = d\tau(1 - v^2/c^2)^{-1/2}$	$dr = dR(1 - 2m/r)^{1/2}$ $dt = d\tau(1 - 2m/r)^{-1/2}$

It is tempting to compare the relations (4.1.5) and (4.1.6) with similar formulae from special relativity (see Table 4.1). However, there are important differences. The square root in the Schwarzschild solution involves the coordinate r and therefore depends on position, whereas that in the special-relativistic case is a constant. Moreover, if we used a different coordinate system for describing the Schwarzschild solution, for example isotropic coordinates (see Problem 3.6), then the expressions would have different forms altogether.

One final point to note is that as $r \to \infty$, $dR \to dr$ in equation (4.1.5) and $d\tau \to dt$ in equation (4.1.6), so asymptotically the coordinate distance dr coincides with the actual distance dR, and the coordinate time dt with the proper time $d\tau$.

Example 1.   If a stick of length 1 m lies radially in the field of a star where m/r is 10^{-2}, what coordinate distance does it take up?

Answer.   From equation (4.1.5), the coordinate distance is

$$\Delta r = (1 - 2m/r)^{1/2} \, \Delta R$$
$$= (1 - 2 \times 10^{-2})^{1/2} \simeq 0.99 \text{ m}.$$

Example 2.   A long stick is lying radially in the field of a spherical object of mass M. If the r-coordinates of its ends are r_1 and r_2 $(r_1 > r_2)$, what is its length?

Answer.   Since the stick is long, we must integrate the length differential dR. This gives the length as

$$\int_{r_2}^{r_1} (1 - 2GM/rc^2)^{-1/2} \, dr = [r^{1/2}(r - 2GM/c^2)^{1/2}$$
$$+ (2GM/c^2) \ln \{r^{1/2} + (r - 2GM/c^2)^{1/2}\}]_{r_2}^{r_1}. \quad (4.1.7)$$

Note that when $GM/rc^2 \ll 1$ this reduces to $r_1 - r_2$.

Exercises 4.1

1. Check the integral (4.1.7).

4.2   Radar sounding

Suppose that an observer is at a fixed point in space in the field
of a massive object, and that directly between him and this object
there is a small body. We can imagine the observer sending radar
pulses in a radial direction towards the body, these pulses being
reflected by it and subsequently received by the observer at some
later time. Let us calculate the time lapse between transmission
and subsequent reception of a radar pulse by the observer.

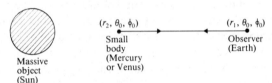

Fig. 4.3   Radar sounding.

If the spatial coordinates of the observer are (r_1, θ_0, ϕ_0), then
those of the body are (r_2, θ_0, ϕ_0), with $r_2 < r_1$ (see Fig. 4.3). The
radar pulses travel in a radial direction with the speed of light, so
putting $d\tau = 0$ and $d\theta = d\phi = 0$ in the line element (4.1.1), we
have

$$(1 - 2m/r)c^2 \, dt^2 = (1 - 2m/r)^{-1} \, dr^2,$$

which gives

$$dr/dt = \pm c(1 - 2m/r).$$

This expression gives the *coordinate speed of light in the radial
direction*. The coordinate time for the whole trip is therefore

$$\Delta t = -\frac{1}{c} \int_{r_1}^{r_2} \frac{dr}{1 - 2m/r} + \frac{1}{c} \int_{r_2}^{r_1} \frac{dr}{1 - 2m/r}$$

$$= \frac{2}{c} \int_{r_2}^{r_1} \frac{dr}{1 - 2m/r}. \tag{4.2.1}$$

However, we require the proper time lapse as measured by the
observer at r_1. (The observer's clock records *proper* time.) From
equation (4.1.6) this is

$$\Delta \tau = \left(1 - \frac{2m}{r_1}\right)^{1/2} \Delta t = \frac{2}{c} \left(1 - \frac{2m}{r_1}\right)^{1/2} \int_{r_2}^{r_1} \frac{dr}{1 - 2m/r}$$

$$= \frac{2}{c} \left(1 - \frac{2m}{r_1}\right)^{1/2} \left(r_1 - r_2 + 2m \ln \frac{r_1 - 2m}{r_2 - 2m}\right). \tag{4.2.2}$$

The distance travelled by the radar pulse is twice the integral (4.1.7), so on the basis of the *classical theory* one would expect a round-tip time of

$$\Delta\tilde{\tau} = (2\times\text{integral } (4.1.7))/c,$$

and $\Delta\tau \neq \Delta\tilde{\tau}$. The difference forms the basis of the so-called fourth test of general relativity, in which the massive object is the Sun, the observer is on Earth, and the small body is either Mercury or Venus. Of course, the Earth is not at a fixed point in space, but we neglect its motion during the travel time of a pulse.

With M equal to the mass of the Sun, and r_1 and r_2 the orbital values of r for the Earth and the other planet involved, $2m/r$ is small for $r_2 < r < r_1$, and this leads to the approximations:

$$\Delta\tau \simeq (2/c)[r_1 - r_2 - m(r_1 - r_2)/r_1 + 2m \ln (r_1/r_2)],$$
$$\Delta\tilde{\tau} \simeq (2/c)[r_1 - r_2 + m \ln (r_1/r_2)]. \tag{4.2.3}$$

Hence there is a general-relativity–induced delay

$$\Delta\tau - \Delta\tilde{\tau} \simeq \frac{2GM}{c^3}\left(\ln \frac{r_1}{r_2} - \frac{r_1 - r_2}{r_1}\right). \tag{4.2.4}$$

For inferior conjunction, with the planet between the Earth and the Sun, this time delay is too small to measure. However, it is increased considerably if they are in superior conjunction, and an experiment was suggested by Shapiro in 1964 which involved radar sounding of Mercury and Venus as they passed behind the Sun [2]. The analysis above will not cope with this situation, where the Sun prevents direct radar sounding in the radial direction.

In using a time-delay formula such as formula (4.2.4), or its modification for non-radial motion, one should ask oneself certain questions. Can the Earth's motion in its orbit be ignored? Can the Earth's own gravitational field be ignored? Can accepted planetary distances be used for r_1 and r_2, which are after all coordinate values and not distances (see Section 4.1)? What is the effect of dispersion by the solar wind? When such considerations have been taken into account, one may go ahead and perform one's experiment to check the theoretical with the observed time delay. Recent tests using Mercury and Venus have yielded agreement to well within the experimental uncertainty of 20 per cent in 1968, and 5 per cent in 1971, while tests using the spacecrafts *Mariner 6* and *7* have yielded agreement to well within the experimental uncertainty of 3 per cent in 1975 [3].

Exercises 4.2

1. Check the approximations (4.2.3).

4.3   Spectral shift

Suppose that a signal is sent from an emitter at a fixed point (r_E, θ_E, ϕ_E), that it travels along a null geodesic and is received by a receiver at a fixed point (r_R, θ_R, ϕ_R). If t_E is the coordinate time of emission and t_R the coordinate time of reception, then the signal passes from the event with coordinates $(t_E, r_E, \theta_E, \phi_E)$ to that with coordinates $(t_R, r_R, \theta_R, \phi_R)$ (see Fig. 4.4). Let u be an affine parameter along the null geodesic with $u = u_E$ at the event of emission and $u = u_R$ at the event of reception. Since the geodesic is null,

$$(1-2m/r)c^2 \, (\mathrm{d}t/\mathrm{d}u)^2 = (1-2m/r)^{-1} \, (\mathrm{d}r/\mathrm{d}u)^2 \\ + r^2(\mathrm{d}\theta/\mathrm{d}u)^2 + r^2 \sin^2\theta \, (\mathrm{d}\phi/\mathrm{d}u)^2,$$

so

$$\frac{\mathrm{d}t}{\mathrm{d}u} = \frac{1}{c}\left[\left(1-\frac{2m}{r}\right)^{-1} \tilde{g}_{ij}\frac{\mathrm{d}x^i}{\mathrm{d}u}\frac{\mathrm{d}x^j}{\mathrm{d}u}\right]^{1/2} \qquad (4.3.1)$$

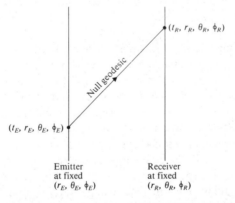

Fig. 4.4   Spacetime diagram illustrating emission and reception of a signal.

where $\tilde{g}_{ij} = -g_{ij}$. On integrating we have

$$t_R - t_E = \frac{1}{c} \int_{u_E}^{u_R} \left[\left(1 - \frac{2m}{r} \right)^{-1} \tilde{g}_{ij} \frac{dx^i}{du} \frac{dx^j}{du} \right]^{1/2} du.$$

The integral on the right-hand side depends only on the path through space, so with a spatially fixed emitter and a spatially fixed receiver, $t_R - t_E$ is the same for all signals sent. So for two signals we have

$$t_R^{(1)} - t_E^{(1)} = t_R^{(2)} - t_E^{(2)},$$

giving

$$\Delta t_R \equiv t_R^{(2)} - t_R^{(1)} = t_E^{(2)} - t_E^{(1)} \equiv \Delta t_E. \tag{4.3.2}$$

That is, the coordinate time difference at the point of emission equals the coordinate time difference at the point of reception. However, the clock of an observer situated at the point of emission records proper time and not coordinate time, the two being related by the finite version of equation (4.1.6). This gives a proper time interval

$$\Delta\tau_E = (1 - 2m/r_E)^{1/2} \Delta t_E,$$

and similarly,

$$\Delta\tau_R = (1 - 2m/r_R)^{1/2} \Delta t_R.$$

Since $\Delta t_R = \Delta t_E$, we have

$$\frac{\Delta\tau_R}{\Delta\tau_E} = \left[\frac{1 - 2m/r_R}{1 - 2m/r_E} \right]^{1/2} \tag{4.3.3}$$

Equation (4.3.3) is the basis of the gravitational spectral-shift formula, which we shall now derive.

Suppose the emitter is a pulsating atom, and that in the proper time interval $\Delta\tau_E$ it emits n pulses. An observer situated at the emitter will assign to the atom a frequency of pulsation $\nu_E \equiv n/\Delta\tau_E$, and this is the *proper frequency* of the pulsating atom. An observer situated at the receiver will see these n pulses in a proper time interval $\Delta\tau_R$ (see Fig. 4.5), and therefore assign a frequency $\nu_R \equiv n/\Delta\tau_R$ to the pulsating atom. Since $\Delta\tau_R \neq \Delta\tau_E$ the observed frequency differs from the proper frequency. In fact,

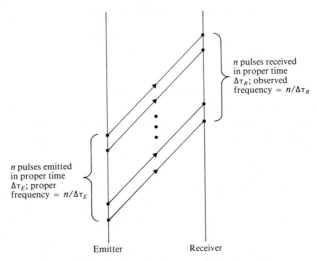

Fig. 4.5   Proper and observed frequencies.

equation (4.3.3) gives

$$\frac{\nu_R}{\nu_E} = \left[\frac{1-2m/r_E}{1-2m/r_R}\right]^{1/2} = \left[\frac{1-2GM/r_Ec^2}{1-2GM/r_Rc^2}\right]^{1/2}, \qquad (4.3.4)$$

on putting $m \equiv GM/c^2$. If $r_Ec^2 \gg 2GM$ and $r_Rc^2 \gg 2GM$, then this reduces to

$$\frac{\nu_R}{\nu_E} \simeq 1 + \frac{GM}{c^2}\left(\frac{1}{r_R} - \frac{1}{r_E}\right). \qquad (4.3.5)$$

From this we can obtain the fractional shift

$$\frac{\Delta\nu}{\nu_E} \equiv \frac{\nu_R - \nu_E}{\nu_E} \simeq \frac{GM}{c^2}\left(\frac{1}{r_R} - \frac{1}{r_E}\right). \qquad (4.3.6)$$

If the emitter is nearer the massive object than the receiver is, then $1/r_R < 1/r_E$, and the shift is towards the red, but if the receiver is nearer the massive object, then it is towards the blue.

There are two points relating to the formulae of spectral shift which are worth noting. The first is that the formula (4.3.4) generalises to any static spacetime, i.e. one for which there exists a timelike coordinate t which gives a splitting of the line element into the form

$$c^2 d\tau^2 = g_{00}(x^k) dt^2 + g_{ij}(x^k) dx^i dx^j.$$

In such a spacetime it makes sense to talk of fixed points in space, and following an argument analogous to that above leads to

$$\nu_R/\nu_E = [g_{00}(x_E^k)/g_{00}(x_R^k)]^{1/2}, \qquad (4.3.7)$$

where x_E^k are the spatial coordinates of the emitter and x_R^k are those of the receiver.

The second point is that the version (4.3.6) may be derived using an eclectic argument not based on general relativity. Suppose for the sake of argument that the emitter is nearer the massive object than the receiver is. Then in travelling from the emitter to the receiver a photon suffers a loss in "intrinsic" energy equal to its gain in gravitational potential energy. The loss in "intrinsic" energy is $h(\nu_E - \nu_R)$, while the gain in gravitational potential energy is

$$\frac{h\nu_E GM}{c^2}\left(\frac{1}{r_E} - \frac{1}{r_R}\right),$$

on assigning the mass $h\nu_E/c^2$ to the photon. Equating these leads to the fractional-shift formula (4.3.6). This formula assumes that the photon's energy has both inertial and gravitational mass, and depends in an essential way on the equivalence principle.

Terrestrial experiments confirming the formula (4.3.6) were performed in 1960 by Pound and Rebka using a vertical separation of 22.5 m in the Jefferson Physics Laboratory at Harvard [4]. The formula should also be amenable to testing by observing the spectra of stars. For an observer on Earth M/r_R is negligible, and the effect depends essentially on M/r_E. Since the observed spectrum is that of atoms on the surface of the star, the effect is greatest for dense objects, such as white dwarfs, for which M/r_E is large. However, data concerning stellar masses and radii are not usually accurate enough for such observations to compete with terrestrial ones. Moreover, the random motion of the radiating atoms produces Doppler shifts which broaden the spectral lines, making it difficult to obtain an accurate value for the gravitational shift.

The following worked example illustrates the use of the spectral-shift formula (4.3.4).

Example. The wavelength of a helium–neon laser is measured inside a *Skylab* freely floating far out in deep space, and is found

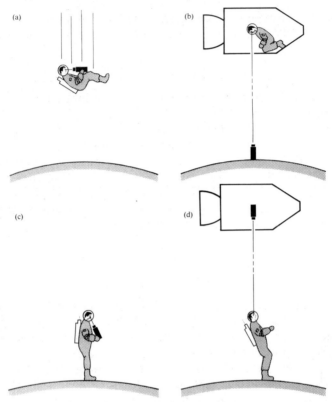

Fig. 4.6   Observer and laser.

to be 632.8 nm. What wavelength would an experimenter measure (see Fig. 4.6) if:

(a) he and the laser fell freely together towards a neutron star?
(b) he remained in the freely floating *Skylab* while the laser transmitted radially from the surface of the neutron star of mass 10^{30} kg and $r_B = 10^4$ m?
(c) he were beside the laser, both on the surface of the neutron star?
(d) he were on the surface of the neutron star while the laser was back in the distant *Skylab*?

Answer

(a) Since the observer is at rest relative to the laser he observes its proper wavelength as determined in the *Skylab*, namely 632.8 nm.

(b) The wavelength version of formula (4.3.4) is

$$\frac{\lambda_R}{\lambda_E} = \left[\frac{1 - 2MG/r_R c^2}{1 - 2MG/r_E c^2}\right]^{1/2}, \qquad (4.3.8)$$

and if we assume that the *Skylab* is so distant that we may take $1/r_R = 0$, and ignore its motion through space, then this gives an observed wavelength of

$$\lambda_R \simeq \lambda_E (1 - 2MG/r_E c^2)^{-1/2} = 685.6 \text{ nm},$$

on putting $\lambda_E = 632.8$ nm, $G = 6.67 \times 10^{-11}$ N m^2 kg^{-2}, $M = M = 10^{30}$ kg, $r^E = r_B = 10^4$ m and $c = 3 \times 10^8$ m s^{-1}.

(c) Here $r_R = r_E$, so formula (4.3.8) gives a measured wavelength of 632.8 nm.

(d) Here we take $1/r_E = 0$ in formula (4.3.8) giving

$$\lambda_R \simeq \lambda_E (1 - 2MG/r_R c^2)^{1/2},$$

where $\lambda_E = 632.8$ nm and $r_R = r_B = 10^4$ m. This gives a measured wavelength of approximately 584 nm.

Exercises 4.3

1. Find the fractional shift in frequency, as measured on Earth, for light from a star of mass 10^{30} kg, assuming that the photons come from just above the star's atmosphere where $r_B = 1000$ km. Would it be easier to detect the shift of radio waves as opposed to violet light?

4.4   General particle motion (including photons)

The paths of particles with mass moving in the vicinity of a spherical massive object are given by the timelike geodesics of spacetime, while the paths of photons are given by the null geodesics. Our plan for this section is to consider first the timelike geodesics, and then to see what modifications are needed for null geodesics.

For a timelike geodesic we may use its proper time τ as an affine parameter. From Section 2.5 we see that the four geodesic equations are given by

$$\frac{\mathrm{d}}{\mathrm{d}\tau}\left(\frac{\partial L}{\partial \dot{x}^\mu}\right) - \frac{\partial L}{\partial x^\mu} = 0, \tag{4.4.1}$$

where

$$L(\dot{x}^\sigma, x^\sigma) \equiv \tfrac{1}{2} g_{\mu\nu} \dot{x}^\mu \dot{x}^\nu$$
$$= \tfrac{1}{2}[c^2(1-2m/r)\dot{t}^2 - (1-2m/r)^{-1}\dot{r}^2 - r^2(\dot{\theta}^2 + \sin^2\theta\dot{\phi}^2)].$$

Here dots denote derivatives with respect to τ, the coordinates are $x^0 \equiv t$, $x^1 \equiv r$, $x^2 \equiv \theta$, $x^3 \equiv \phi$, and we have again put $m = GM/c^2$.

Because of the spherical symmetry, there is no loss of generality in confining our attention to particles moving in the "equatorial plane" given by $\theta = \pi/2$. With this value for θ, the third ($\mu = 2$) of equations (4.4.1) is satisfied, and the second of these ($\mu = 1$) reduces to

$$\left(1-\frac{2m}{r}\right)^{-1}\ddot{r} + \frac{mc^2}{r^2}\dot{t}^2 - \left(1-\frac{2m}{r}\right)^{-2}\frac{m}{r^2}\dot{r}^2 - r\dot{\phi}^2 = 0. \tag{4.4.2}$$

Since t and ϕ are cyclic coordinates, we have immediate integrals of the two remaining equations (see Section 2.5):

$$\partial L/\partial \dot{t} = \text{const}, \qquad \partial L/\partial \dot{\phi} = \text{const}.$$

With $\theta = \pi/2$ these are:

$$(1-2m/r)\dot{t} = k, \tag{4.4.3}$$

$$r^2\dot{\phi} = h, \tag{4.4.4}$$

where k and h are integration constants. We also have the relation (2.6.5) which defines τ. With $\theta = \pi/2$ this becomes

$$c^2(1-2m/r)\dot{t}^2 - (1-2m/r)^{-1}\dot{r}^2 - r^2\dot{\phi}^2 = c^2, \tag{4.4.5}$$

and may often be used in place of the rather complicated equation (4.4.2).

Equation (4.4.3) gives the relation between the coordinate time t and the proper time τ; equation (4.4.4) is clearly analogous to the equation of conservation of angular momentum; as we shall see, equation (4.4.5) yields an equation analogous to that expressing the conservation energy.

Equation (4.4.5) gives

$$c^2(1-2m/r)\dot{t}^2/\dot{\phi}^2 - (1-2m/r)^{-1}(dr/d\phi)^2 - r^2 = c^2/\dot{\phi}^2,$$

and substituting for $\dot{\phi}$ and $\dot{t}$ from equations (4.4.3) and (4.4.4) gives

$$(dr/d\phi)^2 + r^2(1+c^2r^2/h^2)(1-2m/r) - c^2k^2r^4/h^2 = 0.$$

If we put $u \equiv 1/r$ and $m = GM/c^2$ this reduces to

$$\left(\frac{du}{d\phi}\right)^2 + u^2 = E + \frac{2GM}{h^2}u + \frac{2GM}{c^2}u^3, \qquad (4.4.6)$$

where $E \equiv c^2(k^2-1)/h^2$. Comparing this with the analogous Newtonian equation (4.5.1) we see that it corresponds to an energy equation. Comparison also shows that the last term on the right is, in a sense, a relativistic correction, and this is the point of view we shall adopt when discussing the advance of the perihelion in planetary orbits in the next section. In theory equation (4.4.6) may be integrated to give u, and hence r, as a function of ϕ, to obtain the particle paths in the "equatorial plane". Except in special cases, this integration is impossible in practice, and we resort to approximation methods when discussing planetary motion.

Two interesting special cases may be examined in detail, namely vertical free-fall and motion in a circle.

Vertical free-fall. For vertical free-fall, ϕ is constant, so equation (4.4.4) is satisfied with $h = 0$. In deriving equation (4.4.6) we assumed that $\dot{\phi}$ and h were non-zero, so that equation cannot be used. However, it was based on equation (4.4.5), which, with $\dot{\phi} = 0$ and the expression for $\dot{t}$ given by equation (4.4.3) substituted, reduces to

$$\dot{r}^2 - c^2k^2 + c^2(1-2m/r) = 0. \qquad (4.4.7)$$

This equation enables us to give a meaning to the integration constant k, for if the particle is at rest ($\dot{r} = 0$) when $r = r_0$, then $k^2 = 1 - 2m/r_0$. Since τ increases with t, equation (4.4.3) shows that k is the positive square root of $1 - 2m/r_0$ [5]. Hence k is not a universal constant, but depends on the geodesic in question. In particular, if $\dot{r} \to 0$ as $r \to \infty$, then $k = 1$.

Differentiating equation (4.4.7) gives

$$2\dot{r}\ddot{r} + (2mc^2/r^2)\dot{r} = 0,$$

or

$$\ddot{r} + GM/r^2 = 0. \tag{4.4.8}$$

This equation has exactly the same form as its Newtonian counterpart. However, it should be remembered that in equation (4.4.8) the coordinate r is not the vertical distance, and dots are derivatives with respect to proper time, whereas in the Newtonian version r would be vertical distance and dots would be derivatives with respect to the universal time.

Putting $k^2 = 1 - 2m/r_0$ and $m = MG/c^2$ in equation (4.4.7), we get

$$\tfrac{1}{2}\dot{r}^2 = MG\left(\frac{1}{r} - \frac{1}{r_0}\right). \tag{4.4.9}$$

Since the left-hand side is positive, this only makes sense if $r < r_0$. It has exactly the same form as the Newtonian equation expressing the fact that a particle (of unit mass) falling from rest at $r = r_0$ gains a kinetic energy equal to the loss in gravitational potential energy. However, the different meanings of r and the dot mentioned above should be borne in mind.

Equation (4.4.9) allows us to calculate the proper time experienced by the particle in falling from rest at $r = r_0$. If $\tau = 0$ when $r = r_0$, then this time is

$$\tau = \frac{1}{\sqrt{2MG}} \int_r^{r_0} \left(\frac{r_0 r}{r_0 - r}\right)^{1/2} \, dr, \tag{4.4.10}$$

where, because $\dot{r} < 0$, we have taken the negative square root when solving equation (4.4.9) for $dr/d\tau$. The lower limit of integration may be taken down to $2GM/c^2$ (i.e. $2m$) unless the boundary of the massive object is reached first, and as $r \to 2m$ the integral clearly remains finite (see Exercise 4.8.2). However, if one calculates the coordinate time t for falling to $r = 2m \equiv 2GM/c^2$, then one finds it to be infinite. Using $\dfrac{dt}{dr} = \dfrac{dt}{d\tau}\dfrac{d\tau}{dr}$, with

$$\frac{dt}{d\tau} = k/(1 - 2m/r) = (1 - 2m/r_0)^{1/2}/(1 - 2m/r)$$

from equation (4.4.3), and

$$\frac{d\tau}{dr} = -\frac{1}{c\sqrt{2m}}\left(\frac{r_0 r}{r_0 - r}\right)^{1/2}$$

from equation (4.4.9) (with $MG = mc^2$), gives

$$\frac{dt}{dr} = -\frac{1}{c\sqrt{2m}}\frac{r^{3/2}(r_0 - 2m)^{1/2}}{(r-2m)(r_0-r)^{1/2}}, \tag{4.4.11}$$

so the coordinate time to fall from $r = r_0$ to $r = 2m + \varepsilon \,(\varepsilon > 0)$ is

$$t_\varepsilon = \left(\frac{r_0 - 2m}{2mc^2}\right)^{1/2}\int_{2m+\varepsilon}^{r_0}\frac{r^{3/2}\,dr}{(r-2m)(r_0 - r)}.$$

With $2m + \varepsilon < r < r_0$ we have $r > 2m$ and $r_0 - r < r_0$, so

$$t_\varepsilon > \left(\frac{r_0 - 2m}{2mc^2}\right)^{1/2}\frac{(2m)^{3/2}}{r_0^{1/2}}\int_{2m+\varepsilon}^{r_0}\frac{dr}{r-2m}.$$

But

$$\int_{2m+\varepsilon}^{r_0}\frac{dr}{r-2m} = \ln\frac{r_0 - 2m}{\varepsilon} \to \infty \quad \text{as} \quad \varepsilon \to 0,$$

showing that $t_\varepsilon \to \infty$ also. Hence the coordinate time taken to fall to $r = 2m$ is infinite, as asserted.

The way in which the coordinate time t depends on r for a radially falling particle becomes more comprehensive if we compare its *coordinate speed* $v(r)$, defined by $v(r) \equiv |dr/dt|$, with that of a particle falling according to the classical Newtonian theory. For simplicity, let us consider a particle falling from rest at infinity. Letting $r_0 \to \infty$ in equation (4.4.11) gives

$$v(r) = (2mc^2)^{1/2}(r-2m)/r^{3/2},$$

whereas the corresponding classical expression is (with $MG = mc^2$)

$$\bar{v}(r) = (2mc^2)^{1/2}/r^{1/2}.$$

A short calculation shows that as r decreases from infinity $v(r)$ increases until it reaches a maximum value of $2c/3\sqrt{3}$ at $r = 6m$, after which $v(r)$ decreases, and $v(r) \to 0$ as $r \to 2m$. On the other hand, as r decreases, $\bar{v}(r)$ increases, and $\bar{v}(r) \to \infty$ as $r \to 0$. The graphs of $v(r)$ and $\bar{v}(r)$ are given in Fig. 4.7.

Motion in a circle. For circular motion in the "equatorial plane"

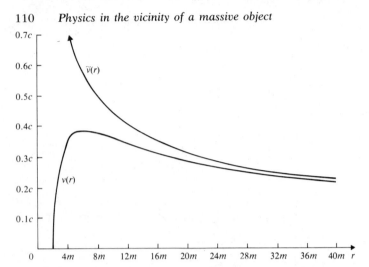

Fig. 4.7 Comparison of the coordinate speed $v(r)$ with the Newtonian speed $\tilde{v}(r)$ for a particle falling from rest at infinity.

we have $r = $ constant, and $\dot{r} = \ddot{r} = 0$. Equation (4.4.2) then reduces to

$$mc^2 \dot{t}^2 = r^3 \dot{\phi}^2, \qquad (4.4.12)$$

giving

$$(d\phi/dt)^2 = GM/r^3, \qquad (4.4.13)$$

on putting $mc^2 = GM$. Hence the change in coordinate time t for one complete revolution is

$$\Delta t = 2\pi (r^3/GM)^{1/2}. \qquad (4.4.14)$$

This expression is exactly the same as the Newtonian expression for the period of a circular orbit of radius r, i.e. Kepler's third law. Although we cannot say that r is the radius of the orbit in the relativistic case, we see that the spatial distance travelled in one complete revolution is $2\pi r$, just as in the Newtonian case.

Figure 4.8 is a spacetime diagram illustrating one complete revolution as viewed by an observer fixed at a point where $r = r_0$. B_1 is the event of the observer's viewing the start of the orbit at A_1, while B_2 is that of his viewing its completion at A_2. The coordinate time between A_1 and A_2 is Δt as given by equation (4.4.14), and we know from the argument used in deriving the spectral-shift formula (see Section 4.3) that Δt is also the coordinate time between B_1 and B_2. So the proper time $\Delta\tau_0$ which the

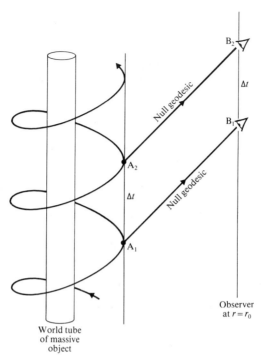

Fig. 4.8 Spacetime diagram illustrating a circular orbit as viewed by a fixed observer.

observer measures for the orbital period is (from equation (4.1.6))

$$\Delta\tau_0 = (1 - 2m/r_0)^{1/2} \, \Delta t. \qquad (4.4.15)$$

As $r_0 \rightarrow \infty$, $\Delta\tau_0 \rightarrow \Delta t$, so Δt is, as it were, the orbital period as measured by an observer at infinity. So, in a sense, Δt is directly observable, and this suggests an indirect means of measuring the coordinate r, by measuring the orbital period of a test particle in a circular orbit given by the value of r. However, this depends on a knowledge of M, which must be known independently, i.e. not found by methods involving orbital periods.

Equation (4.4.15) shows that for a fixed observer the period $\Delta\tau_0$ assigned to the orbit depends on his position. It is natural to ask what period $\Delta\tau$ an observer travelling with the orbiting particle would assign to the orbit. The relationship between t and τ is given by equation (4.4.3), so the answer to the question depends on the value of the integration constant k. From equations

(4.4.12) and (4.4.3) we have

$$\dot{t}^2 = \frac{r^2 k^2}{(r-2m)^2} \quad \text{and} \quad \dot{\phi}^2 = \frac{mc^2 k^2}{r(r-2m)^2}, \qquad (4.4.16)$$

and substitution in equation (4.4.5) (with $\dot{r} = 0$) gives

$$k^2 = \frac{(r-2m)^2}{r(r-3m)}. \qquad (4.4.17)$$

From equation (4.4.3), the orbiting observer assigns a period $\Delta\tau$ to the orbit given by

$$\Delta\tau = (1 - 2m/r)k^{-1}\,\Delta t$$
$$= \left(\frac{r-3m}{r}\right)^{1/2}\Delta t = 2\pi\left[\frac{r^3}{GM}\left(1 - \frac{3MG}{rc^2}\right)\right]^{1/2} \qquad (4.4.18)$$

Since $k^2 > 0$, equation (4.4.17) implies that *circular orbits are impossible, unless $r > 3m$*. In the limit, as $r \to 3m$, $\Delta\tau \to 0$, suggesting that photons can orbit at $r = 3m$, and we shall see later in this section that this is indeed the case.

Imagine now a situation where we have two astronauts in a spacecraft which is in a circular orbit at a value of r greater than $3m$. Suppose one of them leaves the craft, uses his rocket pack to maintain a hovering position at a fixed point in space, and then rejoins the craft after it has completed one orbit. According to equation (4.4.15), the hovering astronaut measures the time of absence as

$$\Delta\tau_{\text{hov}} = (1 - 2m/r)^{1/2}\,\Delta t,$$

while the orbiting astronaut measures it as

$$\Delta\tau_{\text{orb}} = (1 - 2m/r)k^{-1}\,\Delta t,$$

so

$$\frac{\Delta\tau_{\text{hov}}}{\Delta\tau_{\text{orb}}} = \frac{k}{(1 - 2m/r)^{1/2}} = \left(\frac{r-2m}{r-3m}\right)^{1/2} > 1.$$

This shows that if the two astronauts were the same age at the time one of them left the spacecraft for a period of powered flight, then on his return he is older than his companion who remained in the freely falling spacecraft. This result contrasts with the twin paradox of special relativity where the twin undertaking a powered excursion returns to find himself the younger [6].

Photons.   Let us now look at the null geodesics which give the paths of photons (and any other particles having rest mass equal to zero). We cannot use proper time τ as a parameter, so let w be any affine parameter along the geodesic, and let dots now denote derivatives with respect to w. For photons moving in the "equatorial plane", equations (4.4.2) to (4.4.4) remain the same, but the right-hand side of equation (4.4.5) must be replaced by zero:

$$c^2(1-2m/r)\dot{t}^2 - (1-2m/r)^{-1}\dot{r}^2 - r^2\dot{\phi}^2 = 0. \qquad (4.4.19)$$

This leads to a modified form of equation (4.4.6):

$$\left(\frac{\mathrm{d}u}{\mathrm{d}\phi}\right)^2 + u^2 = F + \frac{2GM}{c^2}u^3, \qquad (4.4.20)$$

where $F \equiv c^2 k^2/h^2$. We make use of this equation when discussing the bending of light in Section 4.6. To complete the present section we shall discuss two consequences of the null geodesic equations.

The first is the possibility of having photons in a circular orbit. With $\dot{r} = \ddot{r} = 0$, equation (4.4.2) gives $\dot{\phi}^2/\dot{t}^2 = mc^2/r^3$, while equation (4.4.19) gives $\dot{\phi}^2/\dot{t}^2 = c^2(1-2m/r)/r^2$. Equating these gives $r = 3m$ as the only possible value of r for which photons can go into orbit.

The second consequence is that by investigating the radial null geodesics we can discover what sort of picture a fixed observer gets of any particle falling into a black hole. Suppose the observer is fixed at $r = r_0$, and he drops a particle from rest. According to the discussion above it takes an infinite coordinate time to fall to $r = 2m$, and we see from equation (4.4.11) that $\mathrm{d}r/\mathrm{d}t \to 0$ as $r \to 2m$ (see also Fig. 4.7). So in an r, t-diagram the path of the falling particle is asymptotic to $r = 2m$, as shown in Fig. 4.9. However, as we have seen, the proper time to fall to $r = 2m$ as measured by an observer falling with the particle is finite. Moreover, as $r \to 2m$ equation (4.4.9) shows that $\mathrm{d}r/\mathrm{d}\tau \to -c(1-2m/r_0)^{1/2}$, so the particle has not run our of steam by the time it gets down to $r = 2m$ and presumably passes beyond this threshold. (So in some respects our r, t-diagram is misleading.)

What the observer sees is governed by the outgoing radial null geodesics issuing from the falling particle. For such a geodesic,

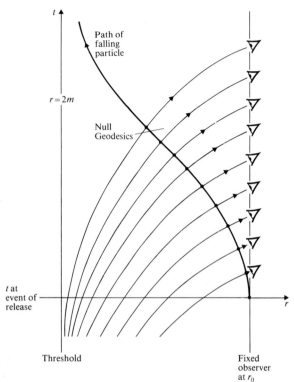

Fig. 4.9   r,t-diagram illustrating the observation of a falling particle by a fixed observer.

equation (4.4.19) gives

$$c^2(1-2m/r)\dot{t}^2 - (1-2m/r)^{-1}\dot{r}^2 = 0,$$

so

$$dr/dt = c(1-2m/r). \tag{4.4.21}$$

As $r \to 2m$, $dr/dt \to 0$, and an outgoing radial null geodesic is also asymptotic to $r = 2m$ in our r, t-diagram. The outgoing null geodesics are therefore as shown in Fig. 4.9. If we follow these back from the eye of the fixed observer, then we discover that he always sees the falling particle *before* it gets to $r = 2m$, as we asserted above.

Our present coordinate system is inadequate for discussing what happens at and beyond $r = 2m$, so care should be exercised in interpreting diagrams such as Figure 4.9. When we discuss black holes in Section 4.8 we make use of another coordinate system which allows us to step over the threshold at $r = 2m$, and to give an improved version of Fig. 4.9.

Exercises 4.4

1. Obtain the second and third of equations (4.4.1) ($\mu = 1$ and 2), and hence show that $\theta = \pi/2$ satisfies the third equation, and that with $\theta = \pi/2$ the second reduces to equation (4.4.2).
2. Check equation (4.4.6) for timelike goedesics, and the corresponding equation (4.4.20) for null geodesics.
3. An observer stationed where $r = r_0$ watches a light signal emitted from a point where $r = r_1$. It travels radially inwards and is reflected by a fixed mirror at $r = r_2$, so that it travels back to its point of origin at $r = r_1$. How long does the round-trip take according to the observer at $r = r_0$? (Assume that $2m < r_2 < r_1 < r_0$.)

4.5 Perihelion advance

For a particle moving in the equatorial plane under the Newtonian gravitational attraction of a spherical object of mass M situated at the origin, classical angular momentum and energy considerations lead to the equation

$$(du/d\phi)^2 + u^2 = E + 2GMu/h^2, \qquad (4.5.1)$$

where $u \equiv 1/r$, E is a constant related to the energy of the orbit, and h is the angular momentum per unit mass given by $r^2\, d\phi/dt = h$ (see Problem 4.6). The solution of this equation is well known from mechanics as

$$u = (GM/h^2)[1 + e \cos(\phi - \phi_0)], \qquad (4.5.2)$$

where ϕ_0 is a constant of integration, and $e \equiv 1 + Eh^4/G^2m^2$. Equation (4.5.2) is that of a conic section with eccentricity e.

The general-relativistic analogue of equation (4.5.1) is equation (4.4.6), and we expect the extra term (equal to $2GMu^3/c^2$) to perturb the Newtonian orbit in some way. If we take the Schwarzschild solution as a model for the solar system, treating the planets as particles, then this extra term makes its presence felt by an advance of the perihelion (i.e. the point of closest approach to the Sun) in each circuit of a planet about the Sun. In deriving this result we make use of an argument due to Møller [7].

The quantity $2GM/c^2$ has the dimensions of length, and is small when compared with values of r corresponding to planetary orbits. Let us denote it by ε, and neglect squares and higher powers of ε. Equation (4.4.6) becomes

$$(\mathrm{d}u/\mathrm{d}\phi)^2 + u^2 = E + (2GM/h^2)u + \varepsilon u^3. \qquad (4.5.3)$$

Aphelion and perihelion occur where $\mathrm{d}u/\mathrm{d}\phi = 0$, i.e. at values of u satisfying

$$\varepsilon u^3 - u^2 + (2GM/h^2)u + E = 0.$$

This is a cubic equation with three roots, u_1, u_2, u_3 say. The sum of the roots is $1/\varepsilon$, which is large. Two of the roots, u_1, u_2 say, are close to their Newtonian counterparts, while the third u_3 is large. Suppose that u_1 gives the aphelion and u_2 the perihelion, so $u_1 \leqslant u \leqslant u_2$. Then

$$\mathrm{d}u/\mathrm{d}\phi = [\varepsilon(u - u_1)(u_2 - u)(u_3 - u)]^{1/2}$$
$$= [(u - u_1)(u_2 - u)]^{1/2}[1 - \varepsilon(u_1 + u_2 + u)]^{1/2},$$

on using $u_1 + u_2 + u_3 = 1/\varepsilon$. So to first order in ε,

$$\frac{\mathrm{d}\phi}{\mathrm{d}u} = \frac{1 + \frac{1}{2}\varepsilon(u_1 + u_2 + u)}{[(u - u_1)(u_2 - u)]^{1/2}}.$$

Putting $\alpha \equiv \frac{1}{2}(u_1 + u_2)$ and $\beta \equiv \frac{1}{2}(u_2 - u_1)$ gives

$$\frac{\mathrm{d}\phi}{\mathrm{d}u} = \frac{\frac{1}{2}\varepsilon(u - \alpha) + 1 + \frac{3}{2}\varepsilon\alpha}{[\beta^2 - (u - \alpha)^2]^{1/2}}.$$

This puts $\mathrm{d}\phi/\mathrm{d}u$ into a form which allows us to integrate to find the angle $\Delta\phi$ between an aphelion and the next perihelion. The answer is

$$\Delta\phi = \left[-\tfrac{1}{2}\varepsilon(\beta^2 - (u - \alpha)^2)^{1/2} + (1 + \tfrac{3}{2}\varepsilon\alpha)\arcsin\frac{u - \alpha}{\beta} \right]_{u_1}^{u_2}$$
$$= (1 + \tfrac{3}{2}\varepsilon\alpha)\pi. \qquad (4.5.4)$$

Doubling $\Delta\phi$ gives the angle between successive perihelions, and shows that in each circuit this is advanced by

$$3\varepsilon\alpha\pi = \frac{3GM\pi}{c^2}(u_1 + u_2) = \frac{3GM\pi}{c^2}\left(\frac{1}{r_1} + \frac{1}{r_2}\right), \qquad (4.5.5)$$

where r_1 and r_2 are the values of r at aphelion and perihelion.

Although the quantity (4.5.5) is incredibly small, the effect is cumulative, and eventually becomes susceptible to observation. It is greatest for the planet Mercury, which is the one closest to the Sun, and amounts to 43″ per century. There is excellent agreement between the theoretical and observed values, but the comparison is not as straightforward as it might seem. In deriving the quantity (4.5.5) we assumed that planets behaved like particles, and ignored their gravitational influence on each other. In fact this influence cannot be ignored, and the effect of the other planets on Mercury causes a perturbation of its orbit. However, after taking this into account (using Newtonian methods) there remains an anomalous advance of the perihelion not explicable in Newtonian terms, and it is this which is accounted for by general relativity. The predicted advances for Mercury, Venus and Earth, together with the observed anomalous advances, are given in the Introduction.

Exercises 4.5

1. Check the calculations leading to the result (4.5.4).

4.6   Bending of light

We have already noted that a massive object can have a considerable effect on light: photons can orbit at $r = 3m$. However, we do not expect to be able to observe this extreme effect in nature, principally because we do not expect to find many objects with $r_B < 3m$. More modest deflections of light passing a massive object can be observed, and in this section we give the theory behind the observations.

The path of a photon travelling in the "equatorial plane" is given by equation (4.4.20). With M turned right down to zero, this becomes

$$(du/d\phi)^2 + u^2 = F, \qquad (4.6.1)$$

a particular solution of which is

$$u = u_0 \sin \phi, \quad \text{or} \quad r_0 = r \sin \phi, \qquad (4.6.2)$$

where $u_0^2 \equiv 1/r_0^2 = F$. This solution represents the straight-line path taken by a photon originating from infinity in the direction $\phi = 0$,

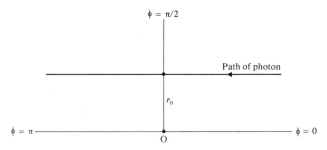

Fig. 4.10   Photon path in the equatorial plane of flat spacetime $(M = 0)$.

and going off to infinity in the direction $\phi = \pi$. The point on the path nearest to the origin O is at a distance r_0 from it, and is given by $\phi = \pi/2$ (see Fig. 4.10). On turning M up, we expect this path to be modified in some way.

Turning M up means replacing equation (4.6.1) by equation (4.4.20). As in the previous section let us put $2GM/c^2 \equiv \varepsilon$, so that equation (4.4.20) becomes

$$(du/d\phi)^2 + u^2 = F + \varepsilon u^3, \qquad (4.6.3)$$

and by confining ourselves to values of r which are large compared with ε, we may treat the second term on the right as a relativistic correction to the flat-spacetime equation (4.6.1), and ignore squares and higher powers of ε. If u_0 is the value of u at the point of closest approach, then $du/d\phi = 0$ when $u = u_0$, so the value of the constant F is $u_0^2(1 - u_0\varepsilon)$, and equation (4.6.3) becomes

$$(du/d\phi)^2 + u^2 = u_0^2(1 - u_0\varepsilon) + \varepsilon u^3. \qquad (4.6.4)$$

This equation should have a solution close to the flat-spacetime solution (4.6.2); let this be

$$u = u_0 \sin \phi + \varepsilon v,$$

where v is some function of ϕ to be determined. Substitution in equation (4.6.4), and working to first order in ε gives

$$2(dv/d\phi) \cos \phi + 2v \sin \phi = u_0^2 (\sin^3 \phi - 1),$$

or

$$d(v \sec \phi)/d\phi = \tfrac{1}{2}u_0^2(\sec \phi \tan \phi - \sin \phi - \sec^2 \phi).$$

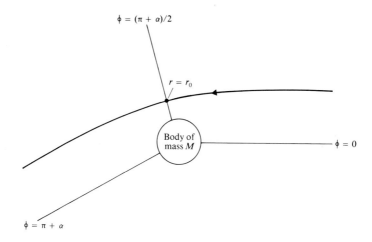

Fig. 4.11 Photon path in the "equatorial plane" of a massive body $(M > 0)$.

Integrating gives

$$v = \tfrac{1}{2}u_0^2(1 + \cos^2 \phi - \sin \phi) + A \cos \phi,$$

where A is a constant of integration. Let us fix A by requiring that the photon originates from infinity in the direction $\phi = 0$, as in the flat-spacetime case. Then $v = 0$ when $\phi = 0$, so $A = -u_0^2$, and

$$u = u_0(1 - \tfrac{1}{2}\varepsilon u_0) \sin \phi + \tfrac{1}{2}\varepsilon u_0^2(1 - \cos \phi)^2 \qquad (4.6.5)$$

is the equation of the path of the photon, to first order in ε.

We no longer expect the photon to go off to infinity in the direction π, but in a direction $\pi + \alpha$, where α is small. Putting $u = 0$ and $\phi = \pi + \alpha$ in equation (4.6.5), and ignoring squares and higher powers of α, and also $\varepsilon\alpha$, gives

$$0 = -u_0\alpha + 2\varepsilon u_0^2,$$

so $\alpha = 2\varepsilon u_0$. (We have used $\sin(\pi + \alpha) \simeq -\alpha$ and $\cos(\pi + \alpha) \simeq -1$). So in its flight past the massive object the photon is deflected through an angle

$$\alpha = 2\varepsilon u_0 = 4GM/r_0 c^2, \qquad (4.6.6)$$

(see Fig. 4.11).

The deflection increases as the impact parameter r_0 decreases.

In the case of light passing through the gravitational field of the Sun, the smallest that r_0 can be is its value at the Sun's surface. If we take for r_0 the accepted value of the Sun's radius (a good enough approximation [8]), then the formula (4.6.6) gives a total deflection (for light originating and terminating at infinity) of 1.75″ [9].

This theoretical result has been checked by observation, but the experiment is a difficult one. One method involves photographing the star-field around the Sun during a total eclipse, and comparing the photograph with one of the same star-field taken six months later. The problems facing experimenters include:

(a) the marked change in conditions which occur when bright sunlight changes to the semi-darkness of an eclipse;
(b) the time-lapse of six months, which makes it difficult to reproduce similar conditions when taking the comparison photograph;
(c) the smallness of the effect, which pushes photography to its limits.

Another method involves the measurement of the relative positions of two radio sources (by interferometric means) as one of them passes behind the Sun [10]. Some detailed figures are given in the Introduction.

4.7   Geodesic effect

If in flat spacetime a spacelike vector λ^μ is transported along a timelike geodesic without rotation (i.e. without changing its spatial orientation), then, in cartesian coordinates, it satisfies $d\lambda^\mu/d\tau = 0$, where τ is the proper time along the geodesic. That is, λ^μ is parallelly transported through *spacetime* along the geodesic. Moreover, if at some point λ^μ is orthogonal to the tangent vector $\dot{x}^\mu \equiv dx^\mu/d\tau$ to the geodesic, then $\eta_{\mu\nu}\lambda^\mu\dot{x}^\nu = 0$, and this relationship is preserved under parallel transport. This orthogonality condition simply means that λ^μ has no temporal component in an instantaneous rest-frame of an observer travelling along the geodesic. The corresponding criteria for transporting a spacelike vector λ^μ in this fashion in the curved spacetime of general relativity are, therefore,

$$d\lambda^\mu/d\tau + \Gamma^\mu_{\nu\sigma}\lambda^\nu\dot{x}^\sigma = 0, \tag{4.7.1}$$

and

$$g_{\mu\nu}\lambda^{\mu}\dot{x}^{\nu} = 0, \tag{4.7.2}$$

where $\dot{x} \equiv dx^{\mu}/d\tau$.

The *geodesic effect* is a consequence of the fact that if a spacelike vector is transported without rotation along a geodesic corresponding to a circular orbit of the Schwarzschild solution, then on its return to the same point in space, after completing one revolution, its spatial orientation has changed. To see this, we must integrate the equations (4.7.1) using expressions for $\Gamma^{\mu}_{\nu\sigma}$ and $\dot{x}^{\sigma}$ corresponding to a circular orbit, which without loss of generality we may take to be in the "equatorial plane".

For such an orbit $\dot{x}^1 = \dot{x}^2 = 0$, and most of the $\Gamma^{\mu}_{\nu\sigma}$ are zero. Making use of the results of Problem 2.7 (with $\theta = \pi/2$), we see that equations (4.7.1) reduce to

$$d\lambda^0/d\tau + \Gamma^0_{10}\lambda^1\dot{x}^0 = 0, \tag{4.7.3a}$$

$$d\lambda^1/d\tau + \Gamma^1_{00}\lambda^0\dot{x}^0 + \Gamma^1_{33}\lambda^3\dot{x}^3 = 0, \tag{4.7.3b}$$

$$d\lambda^2/d\tau = 0, \tag{4.7.3c}$$

$$d\lambda^3/d\tau + \Gamma^3_{13}\lambda^1\dot{x}^3 = 0, \tag{4.7.3d}$$

where

$$\Gamma^0_{10} = \frac{m}{r^2}\left(1 - \frac{2m}{r}\right)^{-1}, \qquad \Gamma^1_{00} = \frac{mc^2}{r^2}\left(1 - \frac{2m}{r}\right),$$

$$\Gamma^1_{33} = -r\left(1 - \frac{2m}{r}\right), \qquad \Gamma^3_{13} = \frac{1}{r}.$$

Let us put $\dot{x}^{\mu} = (a, 0, 0, \Omega a)$, where $a \equiv \dot{t}$ and $\Omega \equiv d\phi/dt$, so that Ω is the angular *coordinate* speed around the circular orbit. Equations (4.4.16), (4.4.17) and (4.4.12) show that

$$a = \left(\frac{r}{r-3m}\right)^{1/2} \quad \text{and} \quad \Omega = c\left(\frac{m}{r^3}\right)^{1/2},$$

on assuming that ϕ increases with t. Both a and Ω are constants. The orthogonality condition (4.7.2) reduces to

$$c^2(1 - 2m/r)\lambda^0\dot{x}^0 - r^2\lambda^3\dot{x}^3 = 0,$$

and allows us to express λ^0 in terms of λ^3:

$$\lambda^0 = [\Omega r^2/c^2(1 - 2m/r)]\lambda^3.$$

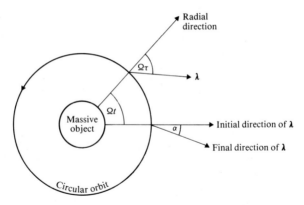

Fig. 4.12   Geodesic effect. Here the initial direction of $\boldsymbol{\lambda}$ ($t = \tau = 0$)
is radial.

A short calculation then shows that equation (4.7.3a) is equivalent to equation (4.7.3d), and that the system of equations
(4.7.3) reduces to

$$\left.\begin{array}{r}
d\lambda^1/d\tau - (r\Omega/a)\lambda^3 = 0, \\
d\lambda^2/d\tau = 0, \\
d\lambda^3/d\tau + (a\Omega/r)\lambda^1 = 0.
\end{array}\right\}\qquad(4.7.4)$$

The general solution of these is

$$\lambda^1 = (A/a)\cos(\phi_0 - \Omega\tau),$$
$$\lambda^2 = B,$$
$$\lambda^3 = (A/r)\sin(\phi_0 - \Omega\tau),$$

where A, B and ϕ_0 are constants of integration. This shows that
the spatial part $\boldsymbol{\lambda}$ of $\lambda^\mu \equiv (\lambda^0, \boldsymbol{\lambda})$ rotates relative to the radial
direction with angular *proper* speed Ω in the negative ϕ-direction.
However, the radial direction itself rotates with angular *coordinate*
speed Ω in the positive ϕ-direction, and it is the difference between angular *proper* speed and angular *coordinate* speed which
gives rise to the geodesic effect.
    If we take the initial direction of $\boldsymbol{\lambda}$ to be radial ($B = \phi_0 = 0$ in
(4.7.5)), and choose the origin of t so that $t = 0$ when $\tau = 0$, then
the situation is as indicated in Fig. 4.12. One revolution is completed in a coordinate time of $2\pi/\Omega$, and hence, from equation
(4.4.18) in a proper time of $\left(\dfrac{r-3m}{r}\right)^{1/2}\dfrac{2\pi}{\Omega}$. The final direction of

$\boldsymbol{\lambda}$ is therefore $2\pi - \alpha$, where

$$\alpha = 2\pi[1 - (1 - 3m/r)^{1/2}].$$

So $\alpha \simeq 3\pi m/r$ for small m/r.

The axis of an orbiting gyroscope furnishes us with a spacelike vector which is transported without rotation, so the geodesic effect is, perhaps, susceptible to observation by means of gyroscopes in orbiting satellites. The smaller the value of r, the greater is the effect, and at the time of writing an experiment involving a terrestrial satellite in a low circular orbit is being prepared [11]. the effect, though small, is cumulative, and for a satellite in near-Earth orbit amounts to about 8″ per year, which should be measurable.

Exercises 4.7

1. What does the geodesic effect amount to for the axis of the Earth in its orbit round the Sun?

   (Take $M_\odot = 2 \times 10^{30}$ kg and $r = 1.5 \times 10^{11}$ m,
   $G = 6.67 \times 10^{-11}$ N m^2 kg^{-2}, $c = 3 \times 10^8$ m s^{-1}.)

4.8 Black holes

So far, our discussion of the Schwarzschild solution has been in terms of the coordinates (t, r, θ, ϕ), and we pointed out in Section 4.0 that the lower bound on r was either its value r_B at the boundary of the object, or $2m(= 2GM/c^2)$, depending on which is reached first as r decreases. If $2m$ is reached first, we have a black hole, and this is the situation prevailing in this section. For an object of mass M, $2GM/c^2$ is known as its *Schwarzschild radius*.

In the limit as $r_E \rightarrow 2m$, the spectral-shift formula (4.3.4) produces an infinite redshift. A particle falling radially inwards appears to continue beyond the threshold at $r = 2m$, although, as we have seen, an observer viewing its fall always sees it before it passes the threshold. These two observations suggest that some odd things happen at $r = 2m$. However, the coordinates (t, r, θ, ϕ) are inadequate for discussing what happens at $r = 2m$ and beyond, so we introduce new coordinates which are valid for $r \leqslant 2m$.

Let us keep r, θ, ϕ, but replace t by

$$v \equiv ct + r + 2m \ln{(r/2m - 1)}. \qquad (4.8.1)$$

A short calculation (see Exercise 4.8.1) shows that in terms of v, r, θ, ϕ the line element is

$$c^2 \, d\tau^2 = (1 - 2m/r) \, dv^2 - 2 \, dv \, dr - r^2 \, d\theta^2 - r^2 \sin^2 \theta \, d\phi^2. \qquad (4.8.2)$$

These new coordinates are *Eddington–Finkelstein coordinates*. They are valid for all v, for all $r > r_B$, even if $r_B < 2m$ (because none of the metric tensor components become infinite), and take us over the threshold at $r = 2m$.

From the line element (4.8.2), we see that radial null geodesics ($d\tau = 0$) are given by

$$\left(1 - \frac{2m}{r}\right)\left(\frac{dv}{dr}\right)^2 - 2\frac{dv}{dr} = 0,$$

i.e. by

$$dv/dr = 0, \qquad (4.8.3)$$

or

$$dv/dr = 2/(1 - 2m/r). \qquad (4.8.4)$$

Differentiation of equation (4.8.1) gives

$$\frac{dv}{dr} = c\frac{dt}{dr} + \frac{1}{1 - 2m/r},$$

so $dv/dr = 0$ implies that $c \, dt/dr = -1/(1 - 2m/r)$, which is negative for $r > 2m$, while $dv/dr = 2/(1 - 2m/r)$ gives $c \, dt/dr = 1/(1 - 2m/r)$, which is positive for $r > 2m$. We therefore infer that equation (4.8.3) gives the ingoing null geodesics, while equation (4.8.4) gives the outgoing ones, at least in the region $r > 2m$.

Integration of equation (4.8.3) gives

$$v = A, \qquad A = \text{constant}, \qquad (4.8.5)$$

while integration of equation (4.8.4) gives

$$v = 2r + 4m \ln|r - 2m| + B, \qquad B = \text{constant}. \qquad (4.8.6)$$

Figure 4.13 shows a v, r-diagram of the radial null geodesics. In drawing this diagram we have used oblique axes, so that the

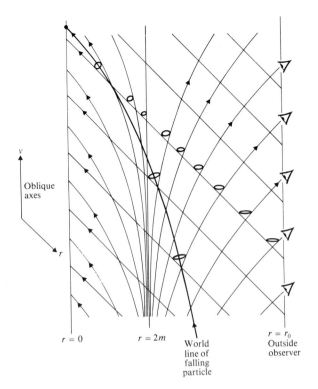

Fig. 4.13 Eddington–Finkelstein picture of ingoing and outgoing null geodesics. Photons fall inwards along lines $v = $ constant, and outwards along curves given by equation (4.8.6).

ingoing null geodesics given by $v = A$ are inclined at 45°, just as they would appear in a flat-spacetime diagram. We have also imagined the whole mass of the object to be concentrated at $r = 0$, and taken the ingoing null geodesics right down to $r = 0$. Equation (4.8.6) shows that $v \to -\infty$ as $r \to 2m$, so the outgoing null geodesics are asymptotic to $r = 2m$, as shown in the figure.

It may be seen from the figure that a photon starting where $r > 2m$ can travel inwards, cross the threshold at $r = 2m$ and carry on inwards, but that a photon starting where $r < 2m$ does not travel outwards. It is confined to the region $r < 2m$. Thus if a massive object had $r_B < 2m$, light could not escape from it to the region $r > 2m$. An outside observer could detect its presence through its gravitational field, but he could not see it, and it is for this reason that such an object is called a *black hole*. Another way of describing the effect of a black hole on light propagation is that it causes the null cones in the tangent spaces to tilt over, and we have drawn small null cones in the figure illustrating this. The possible existence of objects from which light cannot escape was considered as early as 1798 by Laplace [12].

Since we have not changed the coordinate r, the integral (4.4.10), which gives the proper time for a particle to fall inwards from rest at $r = r_0$ remains the same, but we now see that it is valid for $r < 2m$. This integral may be evaluated (see Exercise 4.8.2), and remains finite as its lower limit tends to zero. For example, if $r_0 = 4m$, then the time taken to fall to $r = 2m$ is $\sqrt{2}m(\pi + 2)/c$, while that taken to fall to $r = 0$ is $2\sqrt{2}m\pi/c$ (see Exercise 4.8.3). The world line of a falling particle is also shown in Fig. 4.13, and this results in an improved version of Fig. 4.9.

Thus if Alice were to fall radially down a black hole (rather than a rabbit hole) clutching a clock and a lantern, then she would complete her fall with a finite time on her clock. However, an outside observer would never see her pass beyond $r = 2m$, but she would effectively disappear from view as the light from her lantern became increasingly red-shifted. Once beyond $r = 2m$ she could no longer signal to the outside observer, nor could she return to tell of her experiences.

The above considerations show that an outside observer cannot see events which occur inside the sphere $r = 2m$, and for this reason the sphere, or rather the hypersurface in spacetime which is its time-development, is called an *event horizon.*

Our discussion of the properties of a black hole would be largely academic, unless there were reasons for believing that they might exist in nature. The possibility of their existence arises from the idea of gravitational collapse. If one imagined a very massive object accreting more matter by gravitational attraction, then a stage would be reached where the mutual gravitational attraction between the constituent particles was so great that the internal repulsive forces between them could no longer hold them apart. The whole object would collapse in on itself: nothing could stop this collapse, and the result would be a black hole. Quite general arguments (not based on the spherically symmetric solution of Schwarzschild) exist to show that a collapsing object leads to a singularity in spacetime [13]. If the collapse is spherically symmetric, then the singularity which is the eventual destination of the collapsing material is given by $r = 0$ in the Schwarzschild solution.

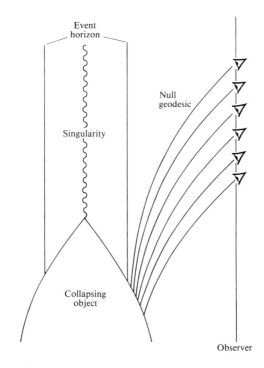

Fig. 4.14 Spacetime diagram of gravitational collapse.

If one assumes that the general features of a collapsing object are not too far removed from those which prevail in the spherically symmetric case, then one would expect the emergence of an event horizon which would shield the object in its collapsed state from view (see Fig. 4.14). An outside observer would see the object to be always outside the event horizon. However, it would effectively disappear from view because of the increasing redshift, and a black hole in space would result. At the time of writing, two possible candidates for black holes have been "observed", namely the compact X-ray source Cygnus X-1 and the nucleus of the radio galaxy M87 [14].

Exercises 4.8

1. Verify the form (4.8.2) of the line element in Eddington–Finkelstein coordinates.

2. By making a substitution $r = r_0 \sin^2 \psi$, show that the value of the integral (4.4.10), giving the proper time to fall radially

from rest at $r = r_0$, is

$$\tau(r_0, r) = \frac{r_0^{3/2}}{c(2m)^{1/2}} \left[\frac{\pi}{2} - \arcsin \left(\frac{r}{r_0} \right)^{1/2} + \left(\frac{r}{r_0} \right)^{1/2} \left(1 - \frac{r}{r_0} \right)^{1/2} \right].$$

3. Using the result of Exercise 4.8.2, show that
   $\tau(4m, 2m) = \sqrt{2}m(\pi + 2)/c$, and that $\tau(4m, r) \to 2\sqrt{2}m\pi/c$ as
   $r \to 0$.

4. Why cannot Alice return to the outside world $(r > 2m)$ after
   falling down the black hole?

4.9   Other coordinate systems

The description of spacetime near a spherically symmetric massive
object need not be in terms of the standard Schwarzschild
coordinates and their corresponding line element. We have
already seen the usefulness of Eddington–Finkelstein coordinates
in discussing what happens beyond the event horizon at $r = 2m$.

Another example is provided by *isotropic coordinates* defined in
Problem 3.6. Here the standard coordinate r is replaced by ρ,
defined by

$$r \equiv \rho(1 + m/2\rho)^2, \tag{4.9.1}$$

and simple substitution gives the line element

$$c^2\, d\tau^2 = c^2(1 - m/2\rho)^2(1 + m/2\rho)^{-2}\, dt^2$$
$$- (1 + m/2\rho)^4\, (d\rho^2 + \rho^2\, d\theta^2 + \rho^2 \sin^2\theta\, d\phi^2). \tag{4.9.2}$$

These coordinates are employed in compiling the relativistic
astronomical tables for the solar system used extensively
throughout the world. We see that the line element has the form

$$c^2\, d\tau^2 = A(\rho)\, dt^2 + B(\rho)\, d\sigma^2,$$

where $d\sigma^2$ is the line element of flat spacetime in spherical polar
coordinates ρ, θ, ϕ.

The particular advantage of the isotropic line element is that
$d\sigma^2$ is invariant under changes of flat-space coordinates, and
ρ, θ, ϕ may therefore be replaced by any other flat-space
coordinates we care to use. For example, if cartesian coordinates
x, y, z (defined in terms of ρ, θ, ϕ) in the usual way) are used as
spatial coordinates, then

$$d\sigma^2 = dx^2 + dy^2 + dz^2,$$

and

$$c^2\, d\tau^2 = A(\rho)\, dt^2 + B(\rho)(dx^2 + dy^2 + dz^2),$$

where ρ occurring in $A(\rho)$ and $B(\rho)$ is given by $\rho^2 = x^2 + y^2 + z^2$.

Our previous results could be formulated in terms of isotropic coordinates, but the corresponding expressions are usually more complicated. For example, corresponding to equation (4.4.21) we would have

$$d\rho/dt = c(1 - m/2\rho)/(1 + m/2\rho)^3. \tag{4.9.3}$$

Kruskal (or *Kruskal–Szekeres*) *coordinates* are, like Eddington–Finkelstein coordinates, particularly useful for discussing what happens both sides of the event horizon. The r and t of the standard Schwarzschild coordinates are replaced by

$$u \equiv (r/2m - 1)^{1/2} e^{r/4m} \cosh(ct/4m),$$
$$v \equiv (r/2m - 1)^{1/2} e^{r/4m} \sinh(ct/4m).$$

This leads to the line element

$$c^2\, d\tau^2 = (32\, m^3/r) e^{-r/2m}(du^2 - dv^2) + r^2(d\theta^2 + \sin^2\theta\, d\phi^2), \tag{4.9.4}$$

where r is defined implicitly by

$$u^2 - v^2 \equiv (r/2m - 1) e^{r/2m}.$$

The particular advantage of these coordinates is that radial null geodesics are given by $u \pm v = $ constant, and are thus straight lines with 45° slopes when drawn in a u, v-diagram, just as in the flat spacetime of special relativity.

We have seen in this chapter the effects on particles in a spacetime which is not flat. Note that no amount of coordinate transformation from one system to another can change the curvature of the spacetime (the test for curvature is given in Section 3.2). Another way of saying this is that we cannot transform away gravity just by turning to another coordinate system, except of course locally, but then only approximately. Globally, we cannot transform away gravity at all.

Exercises 4.9

1. Verify the form (4.9.4) for the line element in Kruskal coordinates.

Problems 4

1. We have referred to the surface in space given by $\theta = \pi/2$ as the "equatorial plane". Show that it is not flat.

2. A free particle of rest mass μ just misses an observer fixed in space where $r = r_0$. Show that he assigns to it an energy

$$E = \mu c^2 (1 - 2m/r_0)^{-1/2}k,$$

where k is the integration constant of equation (4.4.3). (Use Exercise 3.1.1.)

3. Deduce from the previous problem that the "energy at infinity" of a particle is $\mu c^2 k$, and hence that a particle can escape to infinity only if $k \geqslant 1$.

4. Show that the proper time for a photon to complete one revolution at $r = 3m$, as measured by an observer stationed at $r = 3m$, is $6\pi m/c$. What orbital period does a very distant observer assign to the photon?

5. Show, by perturbing the geodesic within the "equatorial plane", that the circular orbit of a photon at $r = 3m$ is unstable.

6. By considering the conservation of energy and angular momentum, show that the path of a particle moving in the equatorial plane under a Newtonian gravitational force due to a spherical object of mass M situated at the origin is given by

$$(du/d\phi)^2 + u^2 = E + 2GMu/h^2,$$

where $u \equiv 1/r$, h is the angular momentum per unit mass and E is a constant.

7. Find the Schwarzschild radius of a spherical object with the same mass as that of the Earth.
   (Take $M_\oplus = 6 \times 10^{24}$ kg, $G = 6.67 \times 10^{-11}$ N m^2 kg^{-2}, $c = 3 \times 10^8$ m s^{-1}.)

8. Suppose we have two spherical objects which are just black holes, i.e. for each, r_B is its Schwarzschild radius. If one has the same mass as the Earth, and the other the same mass as the Sun, which has the greater density?
   (Take $M_\oplus = 6 \times 10^{24}$ kg, $M_\odot = 2 \times 10^{30}$ kg.)

9. What form do equations (4.1.5) and (4.1.6) take in isotropic coordinates?

Notes

1. They could therefore be tests of any other theory of gravitation which yielded the Schwarzschild solution.
2. See Shapiro, 1964.
3. See Shapiro, 1968; Shapiro *et al.*, 1971; and Anderson *et al.*, 1975.
4. See Pound and Rebka, 1960.
5. We are assuming that t has been chosen to increase into the future.
6. This result apparently contradicts the dictum that a timelike geodesic maximises proper time.
7. See Møller, 1972, §12.2.
8. Although r_0 is a coordinate value, the circumference of the Sun's disc is $2\pi r_0$, which is measured by optical means.
9. In 1911, prior to general relativity, Einstein predicted a deflection equal to half this amount. See Hoffmann, 1972, Ch. 8, for history, and Kilmister, 1973, Extract 3, for a translation of Einstein's paper.
10. See, for example, Riley, 1973, where references for other experiments are given.
11. See the paper by Everitt, Fairbank and Hamilton in Carmeli *et al.*, 1970.
12. In classical physics the escape velocity for a particle from a star of mass M and radius r is $(2GM/r)^{1/2}$. Assigning a light corpuscle the escape velocity c yields $r = 2GM/c^2$, which is also the Schwarzschild result. See Hawking and Ellis, 1973, Appendix A, for a translation of Laplace's essay.
13. See, for example, Misner, Thorne and Wheeler, 1973, §34.6.
14. See Thorne, 1974, and Young *et al.*, 1978.

Gravitational radiation

5.0 Introduction

If we ask what characterises radiation, our answers might include the transmission of energy and information through space, or the existence of a wave equation which some quantity satisfies. These aspects are, of course, related, in that there is a characteristic speed of transmission which is determined by the wave equation. In Newtonian gravitational theory energy (and information) is transmitted via the gravitational field which is determined by the gravitational potential V. In empty space V satisfies $\nabla^2 V = 0$, which is not a wave equation, but might be regarded as the limit of a wave equation in which the characteristic speed of transmission tends to infinity. Put another way, gravitational effects are, according to Newton's theory, transmitted instantaneously, which is thoroughly unsatisfactory from the relativistic point of view. Moreover, with an infinite speed of transmission it is impossible to associate a wavelength with a given frequency of oscillation.

Einstein's theory, being a relativistic theory, does not suffer from these defects, and as we shall see, it yields a wave equation for the propagation of gravitational disturbances with a characteristic speed equal to c. A discussion of gravitational radiation using the exact field equations is virtually impossible, because of their extreme non-linearity (although considerable progress has been made in this direction over the last twenty years), and we shall therefore resort to a *linearisation* of the equations appropriate for a weak gravitational field. This leads to the emergence of a wave equation, and allows us to compare gravitational with electromagnetic radiation.

Electromagnetic radiation is generated by accelerating charges, and by analogy we expect accelerating masses to produce gravitational radiation. By the same analogy, we might expect gravitational radiation to be predominantly dipole, but this is not the case. The *mass dipole moment* of a system of particles is, by definition, the 3-vector

$$\mathbf{d} = \sum_{\substack{\text{all} \\ \text{particles}}} m\mathbf{x},$$

where $\mathbf{x}$ is the position vector of a particle of mass m. Hence $\dot{\mathbf{d}}$ is the total momentum of the system, so $\ddot{\mathbf{d}} = \mathbf{0}$ by virtue of conservation of momentum, and it is because of this that we get no dipole radiation. We shall see in Section 5.3 that it is the second time-derivative of the *second moment of the mass distribution* of the source which produces the radiation, showing that it is predominantly *quadrupole* [1]. This second moment is the tensor $\mathbf{I}$ with components defined by [2]

$$I^{ij} = \sum_{\substack{\text{all} \\ \text{particles}}} m x^i x^j$$

(i.e. $\mathbf{I} \equiv \sum m\mathbf{x} \otimes \mathbf{x}$), which for a continuous distribution takes the form of a volume integral

$$I^{ij} \equiv \int \rho x^i x^j \, \mathrm{d}V.$$

We make use of both these forms in our discussion of generation of radiation in Section 5.3.

5.1 What wiggles?

As explained in the previous section, our approach to gravitational radiation is via a linearisation of Einstein's theory appropriate for a weak field. This means that over extensive regions of spacetime there exist nearly cartesian coordinate systems in which

$$g_{\mu\nu} = \eta_{\mu\nu} + h_{\mu\nu}, \tag{5.1.1}$$

where the $h_{\mu\nu}$ are small compared with unity, and such coordinate systems will be used throughout this chapter. The rules of the "linearisation game" are as follows:

(a) $h_{\mu\nu}$ together with its first derivatives $h_{\mu\nu,\rho}$ and higher derivatives are small, and all products of these are ignored;
(b) suffices are raised and lowered using $\eta^{\mu\nu}$ and $\eta_{\mu\nu}$, rather than $g^{\mu\nu}$ and $g_{\mu\nu}$.

The situation is like that of Section 3.4, but *without* the quasi-static condition. As a consequence of (a) and (b), all quantities having the kernel letter h are small, and products of them are ignored. The normal symbol for equality will be used to indicate equality up to first order in small quantities, as well as exact equality.

With these preliminaries explained, we have (see Exercise 2.8.1) $g^{\mu\nu} = \eta^{\mu\nu} - h^{\mu\nu}$, and

$$\Gamma^{\mu}_{\nu\sigma} = \tfrac{1}{2}\eta^{\mu\beta}(h_{\sigma\beta,\nu} + h_{\nu\beta,\sigma} - h_{\nu\sigma,\beta}) = \tfrac{1}{2}(h^{\mu}_{\sigma,\nu} + h^{\mu}_{\nu,\sigma} - h_{\nu\sigma}{}^{,\mu}),$$
(5.1.2)

on putting $\eta^{\mu\beta}h_{\nu\sigma,\beta} = h_{\nu\sigma}{}^{,\mu}$. So the Ricci tensor is

$$R_{\mu\nu} = \Gamma^{\alpha}_{\mu\alpha,\nu} - \Gamma^{\alpha}_{\mu\nu,\alpha} = \tfrac{1}{2}(h_{,\mu\nu} - h^{\alpha}_{\nu,\mu\alpha} - h^{\alpha}_{\mu,\nu\alpha} + h_{\mu\nu,\alpha}{}^{\alpha}), \quad (5.1.3)$$

where $h \equiv h^{\mu}_{\mu} = \eta^{\mu\nu}h_{\mu\nu}$, and the curvature scalar is

$$R \equiv g^{\mu\nu}R_{\mu\nu} = \eta^{\mu\nu}R_{\mu\nu} = h_{,\alpha}{}^{\alpha} - h^{\alpha\beta}{}_{,\alpha\beta}, \quad (5.1.4)$$

on relabelling suffixes. The covariant form of the field equations (3.3.3) then yields

$$h_{,\mu\nu} - h^{\alpha}_{\nu,\mu\alpha} - h^{\alpha}_{\mu,\nu\alpha} + h_{\mu\nu,\alpha}{}^{\alpha} - \eta_{\mu\nu}(h_{,\alpha}{}^{\alpha} - h^{\alpha\beta}{}_{,\alpha\beta}) = 2\kappa T_{\mu\nu},$$

and this simplifies to

$$\bar{h}_{\mu\nu,\alpha}{}^{\alpha} + (\eta_{\mu\nu}\bar{h}^{\alpha\beta}{}_{,\alpha\beta} - \bar{h}^{\alpha}_{\nu,\mu\alpha} - \bar{h}^{\alpha}_{\mu,\nu\alpha}) = 2\kappa T_{\mu\nu}, \quad (5.1.5)$$

on putting

$$\bar{h}_{\mu\nu} \equiv h_{\mu\nu} - \tfrac{1}{2}h\eta_{\mu\nu}. \quad (5.1.6)$$

A further simplification may be effected by means of a gauge transformation, a concept which we now explain.

A *gauge transformation* is a small change of coordinates defined by

$$x^{\mu'} \equiv x^{\mu} + \xi^{\mu}(x^{\alpha}), \quad (5.1.7)$$

where the ξ^{μ} are of the same order of smallness as the $h_{\mu\nu}$. Such a small change of coordinates takes a nearly cartesian coordinate system into one of the same kind. The matrix element $X^{\mu'}_{\nu} \equiv \partial x^{\mu'}/\partial x^{\nu}$ is given by

$$X^{\mu'}_{\nu} = \delta^{\mu}_{\nu} + \xi^{\mu}{}_{,\nu}, \quad (5.1.8)$$

and a straightforward calculation (see Exercise 5.1.3) shows that under a gauge transformation

$$h^{\mu'\nu'} = h^{\mu\nu} - \xi^{\mu,\nu} - \xi^{\nu,\mu}, \tag{5.1.9}$$

$$h' = h - 2\xi^{\mu}{}_{,\mu}, \tag{5.1.10}$$

$$\bar{h}^{\mu'\nu'} = \bar{h}^{\mu\nu} - \xi^{\mu,\nu} - \xi^{\nu,\mu} + \eta^{\mu\nu}\xi^{\alpha}{}_{,\alpha}. \tag{5.1.11}$$

The inverse matrix element $X^{\mu}_{\nu'} \equiv \partial x^{\mu}/\partial x^{\nu'}$ is given by

$$X^{\mu}_{\nu'} = \delta^{\mu}_{\nu} - \xi^{\mu}{}_{,\nu}$$

(see Exercise 5.1.3), so that

$$\bar{h}^{\mu'\alpha'}{}_{,\alpha'} = \bar{h}^{\mu'\alpha'}{}_{,\beta}X^{\beta}_{\alpha'} = \bar{h}^{\mu'\alpha'}{}_{,\beta}\delta^{\beta}_{\alpha} = \bar{h}^{\mu'\alpha'}{}_{,\alpha} = \bar{h}^{\mu\alpha}{}_{,\alpha} - \xi^{\mu}{}_{,\alpha}{}^{\alpha}, \tag{5.1.12}$$

on using equation (5.1.11) and simplifying.

If therefore we choose ξ^{μ} to be a solution of

$$\xi^{\mu}{}_{,\alpha}{}^{\alpha} = \bar{h}^{\mu\alpha}{}_{,\alpha}, \tag{5.1.13}$$

then we have $\bar{h}^{\mu'\alpha'}{}_{,\alpha'} = 0$. In the new coordinate system, each term in the bracketed expression on the left of equation (5.1.5) is separately zero, and, on dropping primes, the equation reduces to

$$\bar{h}_{\mu\nu,\alpha}{}^{\alpha} = 2\kappa T_{\mu\nu}. \tag{5.1.14}$$

This simplified equation is valid whenever $\bar{h}^{\mu\nu}$ satisfies the *gauge condition*

$$\bar{h}^{\mu\alpha}{}_{,\alpha} = 0, \tag{5.1.15}$$

and the above considerations show that we can always arrange for it to be satisfied.

This simplification is an exact parallel of that introduced into electromagnetism by adopting the Lorentz gauge condition (see Section A.8). The quantities $\bar{h}^{\mu\nu}$ correspond to the 4-potential A^{μ}, and the gauge condition (5.1.15) corresponds to the Lorentz gauge condition $A^{\mu}{}_{,\mu} = 0$ (see Exercise A.8.1). A gauge transformation $A_{\mu} \to A_{\mu} - \psi_{,\mu}$ will preserve the Lorentz gauge condition if and only if $\psi_{,\mu}{}^{\mu} = 0$. Correspondingly, as equation (5.1.12) shows, a gauge transformation (5.1.7) will preserve the gauge condition (5.1.5) if and only if

$$\xi^{\mu}{}_{,\alpha}{}^{\alpha} = 0. \tag{5.1.16}$$

Let us introduce the *d'Alembertian* $\Box$ defined by

$$\Box \equiv -\eta^{\alpha\beta}\,\partial_{\alpha}\,\partial_{\beta}, \tag{5.1.17}$$

so that

$$\Box = \partial^2/\partial x^2 + \partial^2/\partial y^2 + \partial^2/\partial z^2 - c^{-2}\,\partial^2/\partial t^2 = \nabla^2 - c^{-2}\,\partial^2/\partial t^2, \quad (5.1.18)$$

on putting $x^0 \equiv ct$, $x^1 \equiv x$, $x^2 \equiv y$, $x^3 \equiv z$. Then for any quantity f,

$$f_{,\alpha}{}^{\alpha} = \eta^{\alpha\beta} f_{,\alpha\beta} = -\Box f,$$

and we see that the results above may be summarised as follows.
The quantities $\bar{h}^{\mu\nu} \equiv h^{\mu\nu} - \frac{1}{2}h\eta^{\mu\nu}$ satisfy

$$\Box\bar{h}^{\mu\nu} = -2\kappa T^{\mu\nu}, \qquad (5.1.19)$$

provided the gauge condition

$$\bar{h}^{\mu\nu}{}_{,\nu} = 0 \qquad (5.1.20)$$

holds. The remaining gauge freedom $x^\mu \to x^\mu + \xi^\mu$ preserves the
gauge condition provided ξ^μ satisfies

$$\Box\xi^\mu = 0. \qquad (5.1.21)$$

Inasmuch as equation (5.1.19) is a wave equation with source
term $-2\kappa T^{\mu\nu} \equiv (16\pi G/c^4)T^{\mu\nu}$, the answer to the question posed
in the section heading is $\bar{h}^{\mu\nu}$, a quantity related to $h_{\mu\nu}$, which
represents a perturbation of the metric tensor $g_{\mu\nu}$ away from the
flat metric tensor $\eta_{\mu\nu}$. In empty spacetime equation (5.1.19) reduces to $\Box\bar{h}^{\mu\nu} = 0$, and we see that gravitational radiation propa-
gates through empty spacetime with the speed of light.

Exercises 5.1

1. Check the expressions (5.1.3), (5.1.4) and (5.1.5) given for the
   Ricci tensor, the curvature scalar and the field equations.
2. If $\bar{h}$ is defined by $\bar{h} \equiv \bar{h}^\mu_\mu$, show that $\bar{h} = -h$, and hence that
   $h_{\mu\nu} = \bar{h}_{\mu\nu} - \frac{1}{2}\bar{h}\eta_{\mu\nu}$.
3. Use the equality $g^{\mu\nu} = \eta^{\mu\nu} - h^{\mu\nu}$ to check equation (5.1.9).
   Deduce equation (5.1.10), where $h' \equiv \eta_{\mu\nu}h^{\mu'\nu'}$, and hence verify
   equation (5.1.11).
   Show also that $X^\mu_{\nu'} = \delta^\mu_\nu - \xi^\mu{}_{,\nu'}$.
4. Equations (5.1.19) and (5.1.20) together imply that $T^{\mu\nu}{}_{,\nu} = 0$.
   Is this consistent with $T^{\mu\nu}{}_{;\nu} = 0$?

5.2   Two polarisations

The simplest sort of solution to the empty-spacetime wave equation $\Box \bar{h}^{\mu\nu} = 0$ is that representing a plane wave, given by

$$\bar{h}^{\mu\nu} = \text{Re}\,[A^{\mu\nu} \exp(ik_\alpha x^\alpha)], \qquad (5.2.1)$$

where $[A^{\mu\nu}]$ is the *amplitude matrix* having constant entries, $k^\mu \equiv \eta^{\mu\alpha} k_\alpha$ is the *wave 4-vector* in the direction of propagation, and Re denotes that we take the real part of the bracketed expression following it. It follows from $\Box \bar{h}^{\mu\nu} = 0$ that k^μ is null, and from the gauge condition (5.1.20) that

$$A^{\mu\nu} k_\nu = 0. \qquad (5.2.2)$$

Since $\bar{h}^{\mu\nu} = \bar{h}^{\nu\mu}$, we see that the amplitude matrix has ten different (complex) entries, but the condition (5.2.2) gives four conditions on these, cutting their number down to six (see Exercise 5.2.1). However, we still have the gauge freedom $x^\mu \to x^\mu + \xi^\mu$ (subject to the condition (5.1.21)), and as we shall see, this may be used to reduce the number still further, so that ultimately there are just *two* entries in the amplitude matrix which may be independently specified. This results in two possible polarisations for plane gravitational waves.

To fix our ideas, let us consider a plane wave propagating in the x^3-direction, so that

$$k^\mu = (k, 0, 0, k) \quad \text{and} \quad k_\mu = (k, 0, 0, -k), \qquad (5.2.3)$$

where $k > 0$. Thus $k = \omega/c$, where ω is the angular frequency. Equation (5.2.2) gives $A^{\mu 0} = A^{\mu 3}$, which implies that all the $A^{\mu\nu}$ may be expressed in terms of A^{00}, A^{01}, A^{02}, A^{11}, A^{12} and A^{22}:

$$[A^{\mu\nu}] = \begin{bmatrix} A^{00} & A^{01} & A^{02} & A^{00} \\ A^{01} & A^{11} & A^{12} & A^{01} \\ A^{02} & A^{12} & A^{22} & A^{02} \\ A^{00} & A^{01} & A^{02} & A^{00} \end{bmatrix}. \qquad (5.2.4)$$

Consider now a gauge transformation generated by

$$\xi^\mu = -\text{Re}\,[i\,\varepsilon^\mu \exp(i\,k_\alpha x^\alpha)],$$

where the ε^μ are constants. This satisfies the condition (5.1.21), as required, and has

$$\xi^\mu{}_{,\nu} = \text{Re}\,[\varepsilon^\mu k_\nu \exp(i\,k_\alpha x^\alpha)]. \qquad (5.2.5)$$

In the new gauge the amplitude matrix is defined by

$$\bar{h}^{\mu'\nu'} = \text{Re}\left[A^{\mu'\nu'} \exp\left(i\, k_{\alpha'} x^{\alpha'}\right)\right],$$

and since $\exp\left(i\, k_{\alpha'} x^{\alpha'}\right)$ differs from $\exp\left(i\, k_{\alpha} x^{\alpha}\right)$ by only a first order quantity, substitution in equation (5.1.11) and using equation (5.2.5) gives

$$A^{\mu'\nu'} = A^{\mu\nu} - \varepsilon^\mu k^\nu - k^\mu \varepsilon^\nu + \eta^{\mu\nu}(\varepsilon^\alpha k_\alpha).$$

If we feed in k^μ from equation (5.2.3) and $A^{\mu\nu}$ from equation (5.2.4), then we obtain

$$\left.\begin{array}{ll}
A^{0'0'} = A^{00} - k(\varepsilon^0 + \varepsilon^3), & A^{1'1'} = A^{11} - k(\varepsilon^0 - \varepsilon^3), \\[4pt]
A^{0'1'} = A^{01} - k\varepsilon^1, & A^{1'2'} = A^{12}, \\[4pt]
A^{0'2'} = A^{02} - k\varepsilon^2, & A^{2'2'} = A^{22} - k(\varepsilon^0 - \varepsilon^3).
\end{array}\right\} \quad (5.2.6)$$

So on taking

$$\varepsilon^0 = (2A^{00} + A^{11} + A^{22})/4k, \qquad \varepsilon^1 = A^{01}/k$$

$$\varepsilon^2 = A^{02}/k, \qquad\qquad\qquad \varepsilon^3 = (2A^{00} - A^{11} - A^{22})/4k,$$

we obtain $A^{0'0'} = A^{0'1'} = A^{0'2'} = 0$ and $A^{1'1'} = -A^{2'2'}$.

On dropping the primes, we see that in the new gauge the amplitude matrix has just four entries, A^{11}, A^{12}, A^{21}, A^{22}, and of these only two may be independently specified, because $A^{11} = -A^{22}$ and $A^{12} = A^{21}$. This new gauge, *which is determined by the wave itself*, is known as the *transverse traceless gauge*, or *TT gauge* for short. In this gauge $\bar{h} \equiv \bar{h}^\mu_\mu = 0$ (because $A^{00} = A^{33} = 0$ and $A^{11} = -A^{22}$), and it follows that $h = 0$ so there is no difference between $h_{\mu\nu}$ and $\bar{h}_{\mu\nu}$ (see Exercise 5.1.2). It is because $h = \bar{h} = 0$ that the gauge is called traceless, and it is because $h_{0\mu} = \bar{h}_{0\mu} = 0$ that it is called transverse. We shall work in the TT gauge for the remainder of this section.

If we introduce two *linear polarisation matrices* $[e_1^{\mu\nu}]$ and $[e_2^{\mu\nu}]$, defined by

$$[e_1^{\mu\nu}] = \begin{bmatrix} 0 & 0 & 0 & 0 \\ 0 & 1 & 0 & 0 \\ 0 & 0 & -1 & 0 \\ 0 & 0 & 0 & 0 \end{bmatrix}, \qquad [e_2^{\mu\nu}] = \begin{bmatrix} 0 & 0 & 0 & 0 \\ 0 & 0 & 1 & 0 \\ 0 & 1 & 0 & 0 \\ 0 & 0 & 0 & 0 \end{bmatrix},$$

$$(5.2.7)$$

we see that the general amplitude matrix is a linear combination of them:

$$A^{\mu\nu} = \alpha e_1^{\mu\nu} + \beta e_2^{\mu\nu}, \tag{5.2.8}$$

where α and β are (complex) constants.

The significance of these matrices may be appreciated by reviewing the analogous situation in electromagnetic radiation, where the plane-wave solution to $\Box A^\mu = 0$ is

$$A^\mu = \text{Re}\,[B^\mu \exp\,(i\,k_\alpha x^\alpha)], \qquad B^\mu = \text{constant}.$$

The Lorentz gauge condition $A^\mu{}_{,\mu} = 0$ implies that $B^\mu k_\mu = 0$, which reduces the number of independent components of the *amplitude vector* B^μ to three. If, as before, we consider a wave propagating in the x^3-direction, so that $k^\mu = (k, 0, 0, k)$, then $B^\mu k_\mu = 0$ implies that $B^0 = B^3$, so

$$B^\mu = (B^0, B^1, B^2, B^0),$$

which is analogous to equation (5.2.4). Changing the gauge by putting $A'_\mu = A_\mu - \psi_{,\mu}$, where

$$\psi = -\text{Re}\,[i\,\varepsilon\,\exp\,(i\,k_\alpha x^\alpha)],$$

preserves the Lorentz gauge condition (because $\Box\psi = 0$), and transforms B^μ to

$$(B')^\mu = B^\mu - \varepsilon k^\mu.$$

So

$$(B')^0 = B^0 - \varepsilon k, \qquad (B')^1 = B^1, \qquad (B')^2 = B^2,$$

which are analogous to equations (5.2.6). If therefore we choose $\varepsilon = B^0/k$, then, on dropping primes, we have $B^0 = 0$, and in the new gauge the amplitude vector has just two components (B^1 and B^2) which may be independently specified. This leads to two *linear polarisation vectors*

$$e_1^\mu = (0, 1, 0, 0) \quad \text{and} \quad e_2^\mu = (0, 0, 1, 0), \tag{5.2.9}$$

and the general amplitude vector is a linear combination of these:

$$B^\mu = \alpha e_1^\mu + \beta e_2^\mu.$$

If $B^\mu = \alpha e_1^\mu$, then the force on a free test charge is in the x^1-direction with a magnitude that varies sinusoidally as the wave passes, causing it to oscillate in the x^1-direction, whereas if $B^\mu = \beta e_2^\mu$ the oscillations take place in the x^2-direction. (These facts may be derived by using equation (A.8.13).) The particular combinations $B^\mu = \alpha(e_1^\mu \pm i\, e_2^\mu)$ give *circularly polarised waves* in which the mutually orthogonal oscillations combine, so that the test charge moves in a circle. The polarisation matrices (5.2.7) have a similar effect on free test particles as the gravitational wave passes, as we shall now show.

A short calculation (see Exercise 5.2.2) shows that in the TT gauge $\Gamma_{00}^\mu = 0$, and this implies that the geodesic equation (2.6.7) is satisfied by $\dot{x}^\mu \equiv dx^\mu/d\tau = c\delta_0^\mu$. (This gives $g_{\mu\nu}\dot{x}^\mu\dot{x}^\nu = c^2$, as required, since in the TT gauge $h_{00} = 0$, so $g_{00} = 1$.) Hence curves having constant spatial coordinates are timelike geodesics, and may be taken as the world lines of a cloud of test particles. It follows that a small spacelike vector $\xi^\mu = (0, \xi^1, \xi^2, \xi^3)$ which gives the spatial separation between two nearby particles of the cloud is constant (see also Problem 5.1). However, this does *not* mean that their spatial separation d is constant, for d is given by

$$d^2 = \tilde{g}_{ij}\xi^i\xi^j,$$

where

$$\tilde{g}_{ij} \equiv -g_{ij} = \delta_{ij} - h_{ij},$$

and the h_{ij} are not constant. If we put

$$\zeta^i \equiv \xi^i + \tfrac{1}{2}h_k^i\xi^k, \qquad (5.2.10)$$

then (to first order in $h_{\mu\nu}$)

$$\delta_{ij}\zeta^i\zeta^j = (\delta_{ij} - h_{ij})\xi^i\xi^j = d^2, \qquad (5.2.11)$$

as a short calculation shows (see Exercise 5.2.3). So ζ^i may be regarded as a faithful position vector giving *correct* spatial separations when contracted with the euclidean metric tensor δ_{ij}.

Note that in the TT gauge $h_i^3 = 0$, so equation (5.2.10) gives $\zeta^3 = \xi^3 = $ constant. Hence if the test-particle separation lies in the direction of propagation of the wave, then it is unaffected by the passage of the wave, showing that a gravitational wave is *transverse*.

Let us now select a particular test particle as a reference particle, and refer the motion of others to it by means of ζ^i, using equation (5.2.10). If $A^{\mu\nu} = \alpha e_1^{\mu\nu}$, with α real and positive for convenience, and $\xi^i = (\xi^1, \xi^2, 0)$, then this equation gives

$$\zeta^i = (\xi^1, \xi^2, 0) - \tfrac{1}{2}\alpha \cos k(x^0 - x^3)(\xi^1, -\xi^2, 0). \qquad (5.2.12)$$

Table 5.1 Effect of a plane wave on a transverse ring of test particles.

Value of $k(x^0 - x^3)$		$2n\pi$	$(2n+\frac{1}{2})\pi$	$(2n+1)\pi$	$(2n+\frac{3}{2})\pi$
Displacement of particles from circular configuration	(a) $A^{\mu\nu} = \alpha e_1^{\mu\nu}$				
(b) $A^{\mu\nu} = \alpha e_2^{\mu\nu}$					

So if we consider those particles which, when $\cos k(x^0 - x^3) = 0$, form a circle with the reference particle as centre, this circle lying in a plane perpendicular to the direction of propagation, then as the wave passes, these particles remain coplanar, and at other times their positions are as shown in row (a) of Table 5.1. All this follows from equation (5.2.12).

If, however, $A^{\mu\nu} = \alpha e_2^{\mu\nu}$, again with α real and positive for convenience, then

$$\zeta^i = (\xi^1, \xi^2, 0) - \tfrac{1}{2}\alpha \cos k(x^0 - x^3)(\xi^2, \xi^1, 0), \qquad (5.2.13)$$

resulting in a sequence of diagrams as shown in row (b) of Table 5.1, which may be obtained from those in row (a) by a 45° rotation [3].

In this way we see how the two polarisations of a plane gravitational wave affect the relative displacements of test particles. As in electromagnetic radiation, we may also have circularly polarised waves in which $A^{\mu\nu} = \alpha(e_1^{\mu\nu} \pm i\, e_2^{\mu\nu})$ see Problem 5.2).

Exercises 5.2

1. Show that equation (5.2.2) implies that each $A^{0\mu}$ may be expressed in terms of the A^{ij} for any (null) k^μ.
2. Show that in the TT gauge associated with any plane wave, $\Gamma^\mu_{00} = 0$ and $\Gamma^\mu_{0\nu} = \tfrac{1}{2}h^\mu_{\nu,0}$.
3. Verify equation (5.2.11).
   (Recall that $\eta_{ij} = -\delta_{ij}$.)
4. Check equations (5.2.12) and (5.2.13).
5. In constructing Table 5.1 we took α to be real and positive. What is the effect of having $\alpha = |\alpha|\, e^{i\theta}$, $\theta \neq 0$?

5.3   Simple generation and detection

Equation (5.1.19) gives the relation between the gravitational radiation, represented by $\bar{h}^{\mu\nu}$, and its source, represented by $T^{\mu\nu}$. The solution of this equation is well known from electromagnetism, and may be expressed as a retarded integral:

$$\bar{h}^{\mu\nu}(x^0, \mathbf{x}) = \frac{\kappa}{2\pi} \int \frac{T^{\mu\nu}(x^0 - |\mathbf{x} - \mathbf{x}'|, \mathbf{x}')}{|\mathbf{x} - \mathbf{x}'|} \, dV'. \tag{5.3.1}$$

Here $\mathbf{x}$ represents the spatial coordinates of the *field point* at which $\bar{h}^{\mu\nu}$ is determined, $\mathbf{x}'$ represents those of a point of the source, and $|\mathbf{x} - \mathbf{x}'|$ is the spatial distance between them. The volume integral is taken over the region of spacetime occupied by the points of the source at the retarded times $x^0 - |\mathbf{x} - \mathbf{x}'|$. This region of spacetime is the intersection of the past half of the null cone at the field point with the world tube of the source [4].

Suppose now that the source is some sort of matter distribution localised near the origin O, and that the source particles have speeds which are small compared with c. If we take our field point at a distance r from O which is large compared to the maximum displacements of the source particles from O, then equation (5.3.1) may be approximated by [5]

$$\bar{h}^{\mu\nu}(ct, \mathbf{x}) = -\frac{4G}{c^4 r} \int T^{\mu\nu}(ct - r, \mathbf{x}') \, dV', \tag{5.3.2}$$

on putting $\kappa = -8\pi G/c^4$ and $x^0 = ct$. This approximation is appropriate for looking at the gravitational radiation in the *far zone* or *wave zone*, and by comparison with electromagnetic theory, we expect that (over not too large regions of space) it looks like a plane wave, in which case the radiative part of $\bar{h}^{\mu\nu}$ is completely determined by its spatial part $\bar{h}^{ij}$, as Exercise 5.2.1 shows. It follows that we need only consider $\int T^{ij} \, dV$ (at the retarded time), a neat expression for which may be obtained as follows.

The stress tensor of the source satisfies the conservation equations $T^{\mu\nu}{}_{,\nu} = 0$ (see Exercise 5.1.4). That is,

$$T^{00}{}_{,0} + T^{0k}{}_{,k} = 0, \tag{5.3.3}$$

$$T^{i0}{}_{,0} + T^{ik}{}_{,k} = 0. \tag{5.3.4}$$

Consider the integral identity

$$\int (T^{ik}x^j)_{,k}\,\mathrm{d}V = \int T^{ik}_{\ ,k}x^j\,\mathrm{d}V + \int T^{ij}\,\mathrm{d}V,$$

where the integrals are taken over a region of space enclosing the source, so that $T^{\mu\nu}=0$ on the boundary of the region. The integral on the left is zero, as may be seen by converting it to a surface integral over the boundary using Gauss's theorem. Hence on using equation (5.3.4),

$$\int T^{ij}\,\mathrm{d}V = -\int T^{ik}_{\ ,k}x^j\,\mathrm{d}V = \int T^{i0}_{\ ,0}x^j\,\mathrm{d}V$$

$$= \frac{1}{c}\frac{\mathrm{d}}{\mathrm{d}t}\int T^{i0}x^j\,\mathrm{d}V.$$

Interchanging i and j and adding gives

$$\int T^{ij}\,\mathrm{d}V = \frac{1}{2c}\frac{\mathrm{d}}{\mathrm{d}t}\int (T^{i0}x^j + T^{j0}x^i)\,\mathrm{d}V. \qquad (5.3.5)$$

But

$$\int (T^{0k}x^ix^j)_{,k}\,\mathrm{d}V = \int T^{0k}_{\ ,k}x^ix^j\,\mathrm{d}V + \int (T^{0i}x^j + T^{0j}x^i)\,\mathrm{d}V,$$

where again the integral on the left vanishes by Gauss's theorem. Hence on using equation (5.3.3) we have

$$\int (T^{0i}x^j + T^{0j}x^i)\,\mathrm{d}V = \frac{1}{c}\frac{\mathrm{d}}{\mathrm{d}t}\int T^{00}x^ix^j\,\mathrm{d}V. \qquad (5.3.6)$$

Combining equation (5.3.5) and (5.3.6) gives

$$\int T^{ij}\,\mathrm{d}V = \frac{1}{2c^2}\frac{\mathrm{d}^2}{\mathrm{d}t^2}\int T^{00}x^ix^j\,\mathrm{d}V.$$

For slowly moving source particles $T^{00}\simeq \rho c^2$, where ρ is the proper density, and equation (5.3.2) yields the approximate expression

$$\bar{h}^{ij}(ct,\mathbf{x}) = -\frac{2G}{c^4 r}\frac{\mathrm{d}^2}{\mathrm{d}t^2}\left[\int \rho x^ix^j\,\mathrm{d}V\right]_{\text{ret}}, \qquad (5.3.7)$$

the notation indicating that the integral is evaluated at the retarded time $t - r/c$. The integrand is recognisable as the second moment of the mass distribution, and indicates the essentially quadrupole nature of gravitational radiation (compare remarks at end of Section 5.0).

As an illustration of the above ideas, let us take as a source a dumb-bell consisting of two particles A and B of equal mass M connected by a light rod of length $2a$, the dumb-bell rotating about the x^3-axis in the positive sense with constant angular speed ω, so that the particles remain in the plane $x^3 = 0$ with the mid-point of the rod at O (see Fig. 5.1). The positions of A and B at any time t may be taken to be

$$x^\mu = \pm(a \cos \omega t, a \sin \omega t, 0),$$

and equation (5.3.7) gives (on replacing the integral by a sum over the two source particles)

$$[\bar{h}^{ij}(ct, \mathbf{x})] = -\frac{4GMa^2}{c^4 r} \frac{d^2}{dt^2} \begin{bmatrix} \cos^2 \omega t & \cos \omega t \sin \omega t & 0 \\ \cos \omega t \sin \omega t & \sin^2 \omega t & 0 \\ 0 & 0 & 0 \end{bmatrix}_{\text{ret}}$$

$$= \frac{8GMa^2\omega^2}{c^4 r} \begin{bmatrix} \cos 2\omega(t - r/c) & \sin 2\omega(t - r/c) & 0 \\ \sin 2\omega(t - r/c) & -\cos 2\omega(t - r/c) & 0 \\ 0 & 0 & 0 \end{bmatrix}. \quad (5.3.8)$$

This clearly represents a gravitational wave of angular frequency 2ω.

For field points near to a point on the x^3-axis where $r \simeq x^3$, we have

$$[\bar{h}^{ij}] \simeq \frac{8GMa^2\omega^2}{c^4 r} \operatorname{Re}\left[(e_1^{ij} - i\, e_2^{ij}) \exp \frac{2 i \omega}{c}(x^0 - x^3) \right], \quad (5.3.9)$$

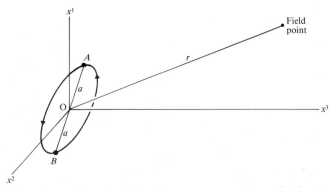

Fig. 5.1   A rotating dumb-bell as a source of gravitational radiation.

and to an observer on the x^3-axis the wave looks like a circularly polarised plane wave (see Section 5.2). Note that this plane-wave approximation automatically has $\bar{h}^{ij}$ in its TT gauge. This does not happen for the plane-wave approximation which an observer at a field point in the plane $x^3 = 0$ holds to be valid, and a transformation to its TT gauge is needed to find its polarisation mode (see Exercise 5.3.1).

Equation (5.3.8) generalises to give the $\bar{h}^{ij}$ produced by a straight bar, having its centre of mass at O and rotating in the plane $x^3 = 0$ with angular speed ω, simply by replacing $2Ma^2$ by the moment of inertia I of the bar about the x^3-axis. It may be shown by methods beyond the scope of this book that the rate at which such an object loses energy by gravitational radiation is given by

$$\frac{\mathrm{d}E}{\mathrm{d}t} = \frac{32GI^2\omega^6}{5c^5}. \tag{5.3.10}$$

This expression is obtained by looking at the energy-momentum carried by the gravitational field itself, which is quadratic in $h_{\mu\nu}$ and its derivatives, and consequently neglected in the linearised theory. As a consequence of the energy loss, ω must decrease, but in the linearised theory it remains constant.

If we feed into the expression (5.3.10) typical laboratory values for I and ω, we find that the power of a laboratory generator is so small that we must look to astrophysical phenomena as possible sources for observable gravitational radiation (see Exercise 5.3.2). These include continuous generators, such as binary stars and pulsating neutron stars, as well as impulsive generators, such as colliding black holes, which would be expected to give off bursts of radiation.

Let us now consider detection. Our discussion of polarisation in Section 5.2 shows that the effect of a gravitational wave on a cloud of free test particles is a variation in their separations; it is as if a varying tidal force were acting on the cloud. If the test particles were not free, but constrained to be the constituent particles of an elastic body, then this tidal force would give rise to vibrations in the body, and here we have the rudiments of a gravitational-wave detector. If the incident radiation were a plane wave of a given frequency, then the responsiveness of such a detector would be enhanced if its fundamental frequency of vibration were to coincide with that of the wave.

Because of the extreme low power of laboratory generators and the extreme distance of astrophysical generators, detection of the predicted radiation was for a long time thought impossible. However, over the past decade or so, considerable ingenuity and effort have gone into the design and construction of detectors, including mechanical ones based on the principle outlined above, and it may not be too long before gravitational astronomy takes its place alongside optical and radio astronomy.

Exercises 5.3

1. Obtain the plane-wave approximation to the radiation from a rotating dumb-bell, held to be valid by an observer at a field point in the plane of rotation, in its TT gauge, and hence show that it is linearly polarised.
   (To relate matters to the theory of Section 5.2, you will find it convenient to put the field point on the x^3-axis, and to let the dumb-bell rotate about one of the other coordinate axes.)
2. Show that the power generated by:
   (a) a steel bar of mass 2×10^5 kg (about 200 tons) and length 10 m, rotating at an angular speed of 50 rad s^{-1} about an axis through its centre of mass which is perpendicular to its length, is about 7.6×10^{-30} watts;
   (b) a binary star, with equal components each of one solar mass, describing circular orbits with a period of one month, is about 5.8×10^{16} watts.
   (Take $G = 6.67 \times 10^{-11}$ N m^2 kg^{-2}, $c = 3 \times 10^8$ m s^{-1}, $M_\odot = 2 \times 10^{30}$ kg.)

Problems 5

1. Show that if in the TT gauge associated with a plane wave $\dot{x}^\mu = c\delta^\mu_0$, then the equation of geodesic deviation (3.3.6) has $\xi^\mu = $ constant as a solution.
2. The separation vector ξ^i of two test particles reacting to a circularly polarised wave propagating in the x^3-direction takes the form $\xi^i = (\xi^1, \xi^2, 0) = $ constant (see Section 5.2 for details). Show that one of the particles moves in a circle with respect to the other.
3. Use equation (5.3.2) to find the $\bar{h}^{\mu\nu}$ in the far zone, due to a single particle of mass M situated at the origin. Obtain the

corresponding line element $c^2 \, d\tau^2 = g_{\mu\nu} \, dx^\mu \, dx^\nu$, and compare it with the approximation of the Schwarzschild line element in isotropic coordinates (equation (4.9.2)) valid for large ρ.

4. Four particles of equal mass are situated at the ends of the arms of a light rigid cross having arms of equal length, and the whole configuration rotates freely about an axis through its centre of mass perpendicular to its plane. Show that in the far zone there is no radiation.

Notes

1. In the classical radiation field associated with a quantum-mechanical particle of integral spin s, the $2s$-pole radiation predominates. Hence gravitons (quadrupole) should have spin 2, just as photons (dipole) have spin 1.
2. Not to be confused with the inertia tensor of Sec. 1.8, which we also denoted by **I**.
3. This expressive means of showing the effect of the polarisation mode on a cloud of test particles is due to Misner, Thorne and Wheeler. See Misner, Thorne and Wheeler, 1973, §35.6.
4. See, for example, Landau and Lifshitz, 1975, §§62, 63.
5. See, for example, Landau and Lifshitz, 1975, §§66, 67. Note that the assumption of small speeds is equivalent to the dimensions of the source being small compared with the wavelength.

Elements of cosmology

6.0 Introduction

The fundamental force keeping solar systems, binary stars and galaxies together is the force of gravity (as opposed to electric, magnetic and nuclear forces), and it is not unreasonable to suppose that the force governing the large-scale motions of the entire universe is primarily gravitational. If there is some other force governing these motions, there has to date been no evidence for it, neither in the solar system nor in the observable galaxies. By the universe we mean all detectable components in the sky: stars, galaxies, constellations, pulsars, quasars, as well as such things as cosmic rays and background radiation. If this directly observable universe is part of a much grander system of universe-within-universes (C.V.I. Charlier's hypothesis [1]), then there is little we can say.

General relativity is a satisfactory theory of gravitation, correctly predicting the motions of particles and photons in curved spacetime, but in order to apply it to the universe we must make some simplifying assumptions. We shall grossly idealise the universe, and model it by a simple macroscopic fluid, devoid of shear-viscous, bulk-viscous and heat-conductive properties. Its stress tensor $T_{\mu\nu}$ is then that of a perfect fluid, so

$$T_{\mu\nu} = (\rho + p)u_\mu u_\nu - p g_{\mu\nu}, \tag{6.0.1}$$

where ρ is its proper density, p is its pressure, u_μ is the (covariant) world velocity of the fluid particles (stars, etc.) and for convenience we have adopted units in which $c = 1$.

Any results we obtain from general relativity should agree with observation. The major items of reliable data that we possess for the universe include the following observed properties:

(i) *Homogeneity*. The number of stars per unit volume, and therefore the density ρ, appear to be uniform throughout large regions of space.

(ii) *Isotropy.* The number of stars per unit solid angle appears to be the same in all directions.

(iii) *Redshift.* There is a redshift $z \equiv \Delta\lambda/\lambda_0$ for the wavelength of light emitted by galaxies, and z increases with distance.

(iv) *Olber's paradox data.* The night sky is not as bright as day. The universe cannot therefore be (spatially) infinite if it is also homogeneous, unless there is a mechanism beyond the inverse-square law for weakening the energy from distant stars and galaxies.

(v) *Background microwave radiation.* Isotropic radiation, apparently corresponding to black-body radiation of about 2.7°K, discovered by Penzias and Wilson in 1965 [2].

(vi) *Ages of meteorites.* Radioactive dating gives the age of meteoric matter as at least 4×10^9 years, and terrestrial matter as about 10^9 years.

In subsequent sections, we shall see how some aspects of the above are incorporated in our treatment of cosmology, but let us first say a little about redshift. If the observed redshift is interpreted as due to a velocity of recession, then the observations may be incorporated in a simple law which states that the speed of recession is proportional to distance. The constant of proportionality is known as *Hubble's constant.* The observations which determine this 'constant' were made over the last few decades, a relatively short period, and there is no reason to believe that it does not change with time. We shall see as the theory develops that it is in fact a function $H(t)$ of time. Its present-day value is estimated to be $(13 \times 10^9)^{-1} \, \mathrm{yrs}^{-1}$.

Exercises 6.0

1. The redshift z is defined by $z \equiv \Delta\lambda/\lambda_0$, where λ_0 is the proper wavelength, and $\Delta\lambda$ is the difference between the observed wavelength and the proper wavelength. If z is a Doppler shift due to a speed of recession v, show that on the basis of special relativity $z \simeq v/c$, for v small compared with c.

6.1 Robertson–Walker line element

It is outside our syllabus to rederive the independent work of Friedmann, Robertson and Walker [3], and others, on metrics,

maximally symmetric subspaces, and descriptions of spacetimes
that comply with the *cosmological principle*. (This is the hypothesis
that the universe is spatially homogeneous and isotropic.) We take
on faith the famous Robertson–Walker line element, adding only
some minor intuitive ideas, and this is the starting point for our
discussions.

With a timelike coordinate t, and spatial coordinates r, θ, ϕ, this
line element is

$$d\tau^2 = dt^2 - [R(t)]^2[(1-kr^2)^{-1}\,dr^2 + r^2\,d\theta^2 + r^2\sin^2\theta\,d\phi^2],$$
$$(6.1.1)$$

where $R(t)$ is a dimensionless scale factor depending only on the
time t, and k is either 0, 1 or -1, and is related to the spatial
curvature. (Again we have taken $c = 1$ for convenience.) The
spatial geometry is determined by the line element

$$ds^2 = (1-kr^2)^{-1}\,dr^2 + r^2\,d\theta^2 + r^2\sin^2\theta\,d\phi^2. \qquad (6.1.2)$$

A 3-dimensional manifold with such a line element is clearly flat
if $k = 0$, but for $k = \pm 1$ it is curved (see Problem 6.1). For $k = 1$
it is a space of constant positive curvature, the 3-dimensional
counterpart of a sphere, and the space is *closed* in the sense that
it has finite volume. For $k = -1$ it is a space of constant negative
curvature, and is *open* in the sense that its volume is infinite. To
justify these remarks would involve us in a lengthy digression into
differential geometry [4]. The scale factor $R(t)$ simply "blows up"
these spaces in a uniform manner, so that they expand or contract
as dR/dt is positive or negative.

The scale factor $R(t)$ operates on the whole spatial part,
regardless of direction, and depends on a *commonly measured
time t*, which we rationalise as follows. Allow each star to carry
its own clock measuring its own proper time τ. These clocks may
(ideally) have been synchronised when $R(t) = 0$, i.e. at the
beginning of the expansion. Because the universe is homogeneous
and isotropic, there is no reason for clocks in different places to
differ in the measurements of their own proper times.
Furthermore, if we tie the coordinate system (t, r, θ, ϕ) to the
stars and galaxies, so that their world lines are given by
$(r, \theta, \phi) = $ constant, then we have a *co-moving coordinate system*
and the time t is nothing other than the proper time τ. This
commonly measured time is often referred to as *cosmic time*.

To gain an intuitive idea of the significance of the Robertson–Walker line element it is useful to imagine a balloon with spots on it to represent the stars and galaxies, the balloon expanding (or contracting) with time. The distance between spots would depend only on a time-varying scale factor $R(t)$, and each spot could be made to possess the same clock-time t. The spatial origin of such a co-moving coordinate system might line on *any one* of the spots.

The line element (6.1.1) is our trial solution for cosmological models, and our next task is to feed it into the field equations (3.3.4) using the form (6.0.1) for $T_{\mu\nu}$. As we shall see, this yields relations involving R, k, ρ and p, and gives a variety of models for comparison with the observed universe. Since our model for the universe assumes homogeneity, ρ and p are functions of t alone.

6.2   Field equations

If we label our coordinates according to $t \equiv x^0$, $r \equiv x^1$, $\theta \equiv x^2$, $\phi \equiv x^3$, then the non-zero connection coefficients are:

$$\left.\begin{array}{lll}
\Gamma^0_{11} = R\dot{R}/(1 - kr^2), & \Gamma^0_{22} = R\dot{R}r^2, & \Gamma^0_{33} = R\dot{R}r^2 \sin^2 \theta, \\
\Gamma^1_{01} = \dot{R}/R, & \Gamma^1_{11} = kr/(1 - kr^2), & \Gamma^1_{22} = -r(1 - kr^2), \\
\Gamma^1_{33} = -r(1 - kr^2) \sin^2 \theta, & & \\
\Gamma^2_{02} = \dot{R}/R, & \Gamma^2_{12} = 1/r, & \Gamma^2_{33} = -\sin \theta \cos \theta, \\
\Gamma^3_{03} = \dot{R}/R, & \Gamma^3_{13} = 1/r, & \Gamma^3_{23} = \cot \theta.
\end{array}\right\}$$

$$(6.2.1)$$

These were obtained in the example in Section 2.5, but we have changed the notation so that derivatives with respect to t are now denoted by dots. Feeding the connection coefficients into

$$R_{\mu\nu} \equiv \Gamma^\sigma_{\mu\sigma,\nu} - \Gamma^\sigma_{\mu\nu,\sigma} + \Gamma^\rho_{\mu\sigma}\Gamma^\sigma_{\rho\nu} - \Gamma^\rho_{\mu\nu}\Gamma^\sigma_{\rho\sigma}$$

(and remembering that $\Gamma^\mu_{\nu\sigma} = \Gamma^\mu_{\sigma\nu}$) gives

$$\left.\begin{array}{l}
R_{00} = 3\ddot{R}/R, \\
R_{11} = -(R\ddot{R} + 2\dot{R}^2 + 2k)/(1 - kr^2), \\
R_{22} = -(R\ddot{R} + 2\dot{R}^2 + 2k)r^2, \\
R_{33} = -(R\ddot{R} + 2\dot{R}^2 + 2k)r^2 \sin^2 \theta, \\
R_{\mu\nu} = 0, \mu \neq \nu.
\end{array}\right\}$$

$$(6.2.2)$$

With $c = 1$, $u^\mu u_\mu = 1$, so

$$T = T^\mu_\mu = (\rho + 4p) - p = \rho - 3p.$$

In our co-moving coordinate system, $u^\mu = \delta^\mu_0$, so

$$u_\mu = g_{\mu\nu}\delta^\nu_0 = g_{\mu 0} = \delta^0_\mu.$$

Hence

$$T_{\mu\nu} = (\rho + p)\delta^0_\mu\delta^0_\nu - pg_{\mu\nu},$$

and

$$\begin{aligned}
T_{\mu\nu} - \tfrac{1}{2}Tg_{\mu\nu} &= (\rho + p)\delta^0_\mu\delta^0_\nu - pg_{\mu\nu} - \tfrac{1}{2}(\rho - 3p)g_{\mu\nu}\\
&= (\rho + p)\delta^0_\mu\delta^0_\nu - \tfrac{1}{2}(\rho - p)g_{\mu\nu}.
\end{aligned}$$

Extracting $g_{\mu\nu}$ from the line element (6.1.1), we see that

$$\begin{aligned}
T_{00} - \tfrac{1}{2}Tg_{00} &= \tfrac{1}{2}(\rho + 3p),\\
T_{11} - \tfrac{1}{2}Tg_{11} &= \tfrac{1}{2}(\rho - p)R^2/(1 - kr^2),\\
T_{22} - \tfrac{1}{2}Tg_{22} &= \tfrac{1}{2}(\rho - p)R^2 r^2,\\
T_{33} - \tfrac{1}{2}Tg_{33} &= \tfrac{1}{2}(\rho - p)R^2 r^2 \sin^2\theta,\\
T_{\mu\nu} - \tfrac{1}{2}Tg_{\mu\nu} &= 0, \qquad \mu \neq \nu.
\end{aligned}$$

So the field equations in the form (3.3.4) yield just two equations:

$$3\ddot{R}/R = \tfrac{1}{2}\kappa(\rho + 3p), \tag{6.2.3}$$

$$R\ddot{R} + 2\dot{R}^2 + 2k = -\tfrac{1}{2}\kappa(\rho - p)R^2, \tag{6.2.4}$$

where (with $c = 1$) $\kappa = -8\pi G$. The fact that the three (non-trivial) spatial equations are equivalent is essentially due to the homogeneity and isotropy of the Robertson–Walker line element.

Eliminating $\ddot{R}$ from equations (6.2.3) and (6.2.4) gives

$$\dot{R}^2 + k = (8\pi G/3)\rho R^2. \tag{6.2.5}$$

We shall refer to this equation as the *Friedmann equation*. Note that the pressure has completely cancelled out of this equation.

We know from Section 3.1 that $T^{\mu\nu}{}_{;\mu} = 0$ yields the continuity equation and the equations of motion of the fluid particles. With $c = 1$ these become (when adapted to curved spacetime)

$$(\rho u^\mu)_{;\mu} + p u^\mu{}_{;\mu} = 0, \tag{6.2.6}$$

$$(\rho + p)u^\nu{}_{;\mu}u^\mu = (g^{\mu\nu} - u^\mu u^\nu)p_{,\mu}. \tag{6.2.7}$$

The continuity equation (6.2.6) may be written as

$$\rho_{,\mu}u^{\mu} + (\rho + p)(u^{\mu}{}_{,\mu} + \Gamma^{\mu}_{\nu\mu}u^{\nu}) = 0,$$

and with $u^{\mu} = \delta^{\mu}_0$ this reduces to

$$\dot{\rho} + (\rho + p)(3\dot{R}/R) = 0, \qquad (6.2.8)$$

which does contain the pressure. As for the equation of motion (6.2.7), both sides turn out to be identically zero, and it is automatically satisfied. This means that the fluid particles (stars, etc.) follow geodesics, which was to be expected, since with p a function of t alone, there is no pressure gradient (i.e. no 3-gradient ∇p) to push them off the geodesics.

We make use of equations (6.2.5) and (6.2.8) in the next section where we discuss the standard Friedmann models of the universe.

Exercises 6.2

1. Check the Ricci tensor components given by equations (6.2.2).
2. Show that equation (6.2.8) may also be derived by eliminating $\ddot{R}$ from equation (6.2.3) and the derivative of equation (6.2.5) with respect to t.
3. Verify that equation (6.2.7) is automatically satisfied.

6.3   The Friedmann models

Observational evidence to date suggests that the universe is matter-dominated, and that the pressure is negligible when compared with the density. The standard Friedmann models arise from setting $p = 0$, and our discussion will be confined to these models only.

With $p = 0$, we see that

$$\rho R^3 = \text{constant} \qquad (6.3.1)$$

is an integral of the continuity equation (6.2.8). As we shall see, this leads to three possible models, each of which has $R(t) = 0$ at some point in time, and it is natural to take this point as the origin of t, so that $R(0) = 0$, and t is then the age of the universe (compare remarks in Section 6.1). Let us use a subscript zero to denote present-day values of quantities, so that t_0 is the present age of the universe, and $R_0 \equiv R(t_0)$ and $\rho_0 \equiv \rho(t_0)$ are the present-day values of R and ρ. We may then write equation (6.3.1) as

$$\rho R^3 = \rho_0 R_0{}^3. \tag{6.3.2}$$

The Friedmann equation (6.2.5) then becomes

$$\dot{R}^2 + k = A^2/R, \tag{6.3.3}$$

where $A^2 \equiv 8\pi G\rho_0 R_0{}^3/3$ $(A > 0)$. *Hubble's "constant"* $H(t)$ is defined by

$$H(t) \equiv \dot{R}(t)/R(t), \tag{6.3.4}$$

and we denote its present-day value by $H_0 \equiv H(t_0)$. Equation (6.2.5) gives

$$\frac{k}{R_0{}^2} = \frac{8\pi G\rho_0}{3} - H_0{}^2 = \frac{8\pi G}{3}\left(\rho_0 - \frac{3H_0{}^2}{8\pi G}\right).$$

Hence $k > 0$, $k = 0$ or $k < 0$ as $\rho_0 > \rho_c$, $\rho_0 = \rho_c$ or $\rho_0 < \rho_c$ respectively, where ρ_c is a *critical density* given by

$$\rho_c \equiv 3H_0{}^2/8\pi G. \tag{6.3.5}$$

The *deceleration parameter* q_0 is defined to be the present-day value of $-R\ddot{R}/\dot{R}^2$. Using equations (6.2.3) (with $p = 0$) and (6.3.4) gives

$$q_0 = 4\pi G\rho_0/3H_0^2 = \rho_0/2\rho_c. \tag{6.3.6}$$

The three Friedmann models arise from integrating equation 6.3.3) for the three possible values of k: $k = 0, \pm 1$.

(i) *Flat model.* $k = 0$; hence $\rho_0 = \rho_c$, $q_0 = \frac{1}{2}$.
    Equation (6.3.3) gives

$$dR/dt = A/R^{1/2},$$

and integrating gives

$$R(t) = (3A/2)^{2/3}t^{2/3}. \tag{6.3.7}$$

This model is also known as the Einstein–de Sitter model, for reasons mentioned in Section 6.4. Its graph is plotted in Fig. 6.1. Note that $\dot{R} \to 0$ as $t \to \infty$.

(ii) *Closed model.* $k = 1$; hence $\rho_0 > \rho_c$, $q_0 > \frac{1}{2}$.

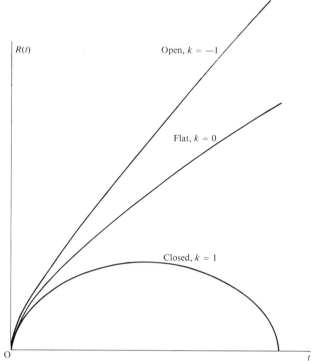

Fig. 6.1   Zero-pressure models of the universe.

Equation (6.3.3) gives

$$\frac{\mathrm{d}R}{\mathrm{d}t} = \left(\frac{A^2 - R}{R}\right)^{1/2},$$

so

$$t = \int_0^R \left(\frac{R}{A^2 - R}\right)^{1/2} \mathrm{d}R.$$

Putting $R \equiv A^2 \sin^2(\psi/2)$ gives

$$t = A^2 \int_0^\psi \sin^2(\psi/2)\,\mathrm{d}\psi = \tfrac{1}{2}A^2 \int_0^\psi (1 - \cos\psi)\,\mathrm{d}\psi$$

$$= \tfrac{1}{2}A^2(\psi - \sin\psi).$$

So

$$R = \tfrac{1}{2}A^2(1 - \cos\psi), \qquad t = \tfrac{1}{2}A^2(\psi - \sin\psi), \qquad (6.3.8)$$

and these two equations give $R(t)$ via the parameter ψ. The graph of $R(t)$ is a cycloid, and is shown in Fig. 6.1.

(iii) *Open model.* $k = -1$; hence $\rho_0 < \rho_c$, $q_0 < \frac{1}{2}$.

Equation (6.3.3) gives

$$\frac{\mathrm{d}R}{\mathrm{d}t} = \left(\frac{A^2 + R}{R}\right)^{1/2},$$

so

$$t = \int_0^R \left(\frac{R}{A^2 + R}\right)^{1/2} \mathrm{d}R.$$

Putting $R \equiv A^2 \sinh^2 (\psi/2)$ gives

$$t = A^2 \int_0^\psi \sinh^2 (\psi/2) \, \mathrm{d}\psi = \tfrac{1}{2}A^2 \int_0^\psi (\cosh \psi - 1) \, \mathrm{d}\psi$$

$$= \tfrac{1}{2}A^2 (\sinh \psi - \psi).$$

So

$$R = \tfrac{1}{2}A^2(\cosh \psi - 1), \qquad t = \tfrac{1}{2}A^2(\sinh \psi - \psi), \qquad (6.3.9)$$

and these give $R(t)$ via the parameter ψ. Its graph is also shown in Fig. 6.1. Note that $\dot{R} = \sinh \psi/(\cosh \psi - 1)$, so $\dot{R} \to 1$ as ψ (and hence $t) \to \infty$.

We see that $k = 0$ and $k = -1$ give models which continually expand, while $k = 1$ gives a model which expands to a maximum value of R, and then contracts, so the latter is not only spatially but also temporally closed. The significance of the value of k is explained in Newtonian terms in Section 6.5.

The question naturally arises as to which (if any, bearing in mind our assumptions) represents our own universe. Astronomical observations over the last forty or so years have yielded estimates for H_0, q_0 and ρ_0. The methods employed are extremely ingenious and complex, and we shall not attempt a review of them [5]. Recent estimates for H_0 put it at about $(13 \times 10)^{-1}$ yrs^{-1}, and using equation (6.3.5) gives a value for the critical density ρ_c of about 1.3×10^{-25} kg m^{-3}. Estimates for the deceleration parameter q_0 are more difficult to obtain, but the evidence suggests that $q_0 \simeq 1$, in which case the universe is closed. However, $q_0 \simeq 1$ gives, via equation (6.3.6),

$$\rho_0 \simeq 2\rho_c \simeq 2.6 \times 10^{-25} \text{ kg m}^{-3},$$

which is much greater than the observed density of the universe estimated on the assumption that its matter content lies principally in the galaxies. This discrepancy leads to the so-called "problem of missing matter" whose existence is postulated to avoid a contradiction. It is suggested that this matter might take the form of an inter-galactic gas, but searches for its existence have proved unsuccessful. On the other hand, if we accept estimates of ρ_0 based on the masses of galaxies and their distribution, this gives $\rho_0 \simeq 0.028\,\rho_c$, $q_0 \simeq 0.014$, and the universe is open. In this case we have a conflict with $q_0 \simeq 1$. The conclusion we come to is not that the Friedmann models are unacceptable, but that we need more observational data and better methods of interpreting them.

One final connection we can make between theory and observation concerns the present age t_0 of the universe. For the flat model, equation (6.3.7) gives

$$H(t) \equiv \dot{R}(t)/R(t) = 2/(3t),$$

so

$$t_0 = 2/(3H_0) \simeq 8.7 \times 10^9 \text{ yrs.}$$

To find the values of t_0 given by the other two models is somewhat complicated, and is left as an exercise (Problem 6.2). With $q_0 \simeq 1$, the closed model gives

$$t_0 \simeq 7.4 \times 10^9 \text{ yrs,}$$

and with $q_0 \simeq 0.014$, the open model gives

$$t_0 \simeq 12.4 \times 10^9 \text{ yrs.}$$

None of these estimates conflicts with, nor is especially supported by, the meteorite data.

Exercises 6.3

1. Check the integrations leading to the parametric equations (6.3.8) and (6.3.9).

6.4   Comment on Einstein's models

As early as 1917 Einstein applied his field equations to a "cosmic gas" of the kind we have been discussing. He was strongly drawn, on philosophical rather than mathematical grounds, to the idea of a stable universe, with $k = 1$, that did not change with time, i.e.

with $H(t) = \dot{R}(t)/R(t) = 0$. In order to counteract the obvious gravitational collapse of such a gas he introduced into the field equations a cosmological term to act as a repulsion mechanism (possibly due to negative matter, if it existed in the universe). The revised field equations were

$$R^{\mu\nu} - \tfrac{1}{2}Rg^{\mu\nu} + \Lambda g^{\mu\nu} = \kappa T^{\mu\nu}, \qquad (6.4.1)$$

where Λ was a constant known as the *cosmological constant*.

Since $g^{\mu\nu}{}_{;\mu} = 0$, this did not alter the divergence property $T^{\mu\nu}{}_{;\mu} = 0$. The constant Λ had to be extremely small, so as not to interfere with the general-relativistic predictions for the solar system (see Chapter 4).

After Hubble's detection of the redshift in the 1920s and 1930s, interpreted now by almost all astronomers and physicists as a Doppler shift due to expansion, Einstein came to believe that the universe was flying apart with considerable kinetic energy, so that a repulsion mechanism was no longer required. He withdrew the cosmological term, and later referred to it as the greatest blunder of his life. In 1932 he proposed with de Sitter a model in which $k = 0$ and $p = 0$, which is the flat model of the previous section. This model leads to a present-day density of $\rho_0 = \rho_c \simeq 1.3 \times 10^{-25}$ kg m^{-3} and to an age $t_0 \simeq 8.7 \times 10^9$ yrs.

Exercises 6.4

1. Show that the empty-spacetime field equations derived from equations (6.4.1) are $R^{\mu\nu} = \Lambda g^{\mu\nu}$.

6.5   Newtonian dust

Suppose we have a Newtonian dust (i.e. a fluid with zero pressure moving according to Newton's laws of motion and gravitation) of uniform density $\rho(t)$, which is in a state of uniform expansion, the only force on it being gravity. This means that the position vector **r** of a fluid particle at any time t is given by a relation of the form

$$\mathbf{r} = R(t)\mathbf{c}, \qquad (6.5.1)$$

where **c** is a constant vector which is determined by the initial position of the fluid particle. Differentiation gives

$$\dot{\mathbf{r}} = H(t)\mathbf{r}, \qquad (6.5.2)$$

where $H(t) = \dot{R}(t)/R(t)$, and in this way our dust has a Hubble 'constant' $H(t)$ associated with the scale factor $R(t)$.

The Newtonian continuity equation

$$\partial \rho/\partial t + \nabla \cdot (\rho \dot{\mathbf{r}}) = 0$$

gives $\dot{\rho} + 3\rho H = 0$, or $R\dot{\rho} + 3\dot{R}\rho = 0$, which integrates to give

$$\rho R^3 = \text{constant}.$$

As in Section 6.3, let us indicate present-day values with a subscript zero, and write the above equation as

$$\rho R^3 = \rho_0 R_0{}^3. \tag{6.5.3}$$

Euler's equation of motion for such a fluid takes the form [6]

$$(\partial/\partial t + \dot{\mathbf{r}} \cdot \nabla)\dot{\mathbf{r}} = \mathbf{F},$$

where $\mathbf{F}$ is the body force per unit mass. With $\dot{\mathbf{r}} = H(t)\mathbf{r}$, this reduces to

$$(\dot{H} + H^2)\mathbf{r} = \mathbf{F}. \tag{6.5.4}$$

The body force $\mathbf{F}$ is due to gravity, and satisfies $\nabla \cdot \mathbf{F} = -4\pi G\rho$, and on taking the divergence of equation (6.5.4) we have

$$3(\dot{H}^2 + H^2) = -4\pi G\rho.$$

Putting $H = \dot{R}/R$ results in

$$3\ddot{R}/R = -4\pi G\rho, \tag{6.5.5}$$

which is exactly the same as the relativistic equation (6.2.3) with p put equal to zero. Substitution for ρ from equation (6.5.3) and rearrangement gives

$$2\ddot{R} + A^2/R^2 = 0,$$

where, as before, $A^2 \equiv 8\pi G\rho_0 R_0{}^3/3$. Multiplying by $\dot{R}$ and integrating gives

$$\dot{R}^2 + k = A^2/R, \tag{6.5.6}$$

where k is a constant of integration. This is exactly the same as the Friedmann equation (6.3.3), but there k is either ± 1 or 0. In fact if $k \neq 0$, there is no loss of generality in taking it to be ± 1 in the Newtonian case (see Exercise 6.5.1), and we are therefore led to exactly the same three models for the evolution of the universe as we obtained in Section 6.3.

Equation (6.5.6) was obtained by integrating equation (6.5.5), which is effectively the equation of motion of the whole dust-filled Newtonian universe. There is therefore a sense in which we may regard equation (6.5.6) as the energy equation of the whole universe. Writing it as

$$\dot{R}^2 - A^2/R = -k.$$

we may regard $\dot{R}^2$ as a measure of its kinetic energy, $-A^2/R$ as a measure of its gravitational potential energy, and $-k$ *as a measure of its total energy*. If $k = -1$, the total energy is positive, and the universe has an excess of kinetic energy which allows it to keep expanding at an ultimately constant rate ($\dot{R} \rightarrow 1$ and $R \rightarrow \infty$ as $t \rightarrow \infty$). If $k = 0$, the total energy is zero, and the kinetic energy is just sufficient to allow the universe to keep expanding, but at a decreasing rate ($\dot{R} \rightarrow 0$ and $R \rightarrow \infty$ as $t \rightarrow \infty$). If $k = 1$, then the universe has insufficient kinetic energy for continued expansion. It expands until $\dot{R} = 0$ (when $R = A^2$) and subsequently contracts. This simplistic treatment of the universe as modelled by a Newtonian dust affords insight into the meaning of the curvature constant k appearing in the relativistic models.

Since the Newtonian analysis leads to the same differential equation and hence the same results as in the relativistic case, we may ask why we bother with a relativistic treatment. Our answer must include the following. In the first place, there are difficulties with Newtonian cosmology which our simple treatment obscures [7]. Secondly, in relativity pressure contributes to the total energy, and hence to the gravitational field, and Newtonian gravity is deficient in this respect. Thirdly, if the fluid contains particles (stars, etc.) having relativistic speeds, then Newtonian physics is inadequate. Finally, the problem of light propagation throughout the universe should be handled from a relativistic viewpoint.

General relativity is essential for a proper treatment of cosmology, though the topics touched on here do not do justice to modern cosmology. A fuller account should include a discussion of luminosities, redshifts, background microwave radiation, galaxy formation and steady-state models, to mention but a few of the topics engaging the attention of astronomers today.

Exercises 6.5

1. The relationship (6.5.1) is preserved if we replace $R(t)$ and $\mathbf{c}$ by $\tilde{R}(t)$ and $\tilde{\mathbf{c}}$, defined by

$$R(t) = \lambda \tilde{R}(t), \qquad \mathbf{c} = \lambda^{-1}\tilde{\mathbf{c}},$$

where λ is constant. Show that this leads to $\tilde{H}(t) = H(t)$, $\tilde{A}^2 = A^2/\lambda^3$, and

$$\dot{\tilde{R}}^2 + k/\lambda^2 = \tilde{A}^2/\tilde{R}$$

in place of equation (6.5.6).
Hence if $k \neq 0$, by a suitable choice of λ we can make $\tilde{k} \equiv k/\lambda^2 = \pm 1$.

Problems 6

1. Show that the 3-dimensional manifold with line element (6.1.2) has a curvature scalar equal to $-6k$.
2. Show that for the cases $k = \pm 1$ the constant of integration A^2 occurring in equation (6.3.3) is given by either

$$\frac{A^2}{2} = R_0 q_0 \left(\frac{k}{2q_0 - 1}\right) \qquad \text{or} \qquad \frac{A^2}{2} = \frac{q_0}{H_0}\left(\frac{k}{2q_0 - 1}\right)^{3/2}.$$

Hence show that for the closed model the present-day value ψ_0 of the parameter ψ (see equations (6.3.8)) is given by $\cos\psi_0 = (1 - q_0)/q_0$, while for the open model (see equations (6.3.9)) it is given by $\cosh\psi_0 = (1 - q_0)/q_0$.
    Use these results to show that if $q_0 \simeq 1$ (closed model), then $t_0 \simeq 7.4 \times 10^9$ yrs, while if $q_0 \simeq 0.014$ (open model), then $t_0 \simeq 12.4 = 10^9$ yrs.
(Take $1/H_0 \simeq 13 \times 10^9$ yrs.)
3. Show that for the flat model the integration constant A^2 of equation (6.3.3) is $A^2 = R_0^3 H_0^2$.
4. Find the radial coordinate speed $|dr/dt|$ of light in a Robertson–Walker universe.
5. In a Robertson–Walker universe, light is emitted from a star with spatial coordinates (r_s, θ_s, ϕ_s). It travels radially inwards, and is received by an observer situated at the origin ($r = 0$). Show that the ratio of the observed wavelength λ to the

proper wavelength λ_0 is given by

$$\lambda/\lambda_0 = R(t_2)/R(t_1),$$

where t_1 is the time of emission, and t_2 is the time of reception.

(Use the result of Problem 4, and neglect any gravitational shift due to the star's own gravitational field.)

6. If in Problem 5 the time difference $\Delta t = t_2 - t_1$ is small, show that the redshift z defined by $1 + z \equiv \lambda/\lambda_0$ is given by $z \simeq H(t_1)\Delta t \simeq H(t_2)\Delta t$. By combining this result with that of Exercise 6.0.1 deduce the relation $v \simeq H(t_2)d$ connecting the speed of recession v with distance d.

7. If the Robertson–Walker universe of Problem 5 is the flat Friedmann model ($k = 0$), show that t_1 and t_2 are connected by

$$r_s = 3(2/3A)^{2/3}(t_2^{1/3} - t_1^{1/3}),$$

where (from Problem 3) $A^2 = R_0^3 H_0^2$. Deduce that a present-day observer can only see those stars for which $r_s \leq 2/R_0 H_0$. (This indicates the presence of a *particle horizon* in the universe.)

Notes

1. See Charlier, 1922.
2. See Penzias and Wilson, 1965.
3. See Friedmann, 1922; Robertson, 1935 and 1936; and Walker, 1936.
4. See, for example, Misner, Thorne and Wheeler, 1973, §27.6, in particular Box 27.2.
5. See, for example, Weinberg, 1972, Ch. 14. The estimates for H_0 and q_0 used in our discussion are taken from Weinberg. More recent observational analysis suggests that H_0 may be as low as $(18 \times 10^9)^{-1}$ yrs^{-1}, and that q_0, based on masses and distributions of galaxies, may be as high as 0.025. Such revisions make no qualitative difference to our discussion.
6. See, for example, Landau and Lifshitz, 1959, §2.
7. See Bondi, 1960, §9.3.

Appendix

Special relativity review

A.0 Introduction

Newton believed that time and space were completely separate entities. Time flowed evenly, the same for everyone, and fixed spatial distances were identical, whoever did the observing. These ideas are still tenable, even for projects like manned rocket travel to the Moon, and almost all the calculations of everyday life in engineering and science rest on Newton's very reasonable tenets. Einstein's 1905 discovery that space and time were just two parts of a single higher entity, *spacetime*, alters only slightly the well-established Newtonian physics with which we are familiar. The new theory is known as *special relativity*, and gives a satisfactory description of all physical phenomena (when allied with quantum theory), with the exception of gravitation. It is of importance in the realm of high relative velocities, and is checked out by experiments performed every day, particularly in high-energy physics. For example, the Stanford linear accelerator, which accelerates electrons close to the speed of light, is about two miles long and cost 10^8; if Newtonian physics were the correct theory, it need only have been about one inch long.

The fundamental postulates of the theory concern *inertial reference systems* or *inertial frames*. Such a reference system is a coordinate system based on three mutually orthogonal basis vectors, which give coordinates x, y, z in space, and an associated system of synchronised clocks at rest in the system, which give a time coordinate t, and which is such that when particle motion is formulated in terms of this reference system *Newton's first law holds*. It follows that if K and K' are inertial frames, then K' is moving relative to K without rotation and with constant velocity. The four coordinates (t, x, y, z) label points in spacetime, and such a point is called an *event*.

The fundamental postulates are:

(i) The speed of light c is the same in all inertial frames.

(ii) The laws of nature are the same in all inertial frames.

Postulate (i) is clearly at variance with Newtonian ideas on light propagation. If the same system of units is used in two inertial frames K and K', then it implies that

$$c = dr/dt = dr'/dt', \qquad (A.0.1)$$

where $dr^2 = dx^2 + dy^2 + dz^2$, and primed quantities refer to the frame K'. Equation (A.0.1) may be written as

$$c^2 dt^2 - dx^2 - dy^2 - dz^2 = c^2(dt')^2 - (dx')^2 - (dy')^2 - (dz')^2 = 0,$$

and is consistent with the assumption that there is an invariant *interval* ds between neighbouring events given by

$$\pm ds^2 = c^2 dt^2 - dx^2 - dy^2 - dz^2$$
$$= c^2(dt')^2 - (dx')^2 - (dy')^2 - (dz')^2, \qquad (A.0.2)$$

which is such that $ds = 0$ for neighbouring events on the spacetime curve representing a photon's history. It is convenient to introduce indexed coordinates x^μ ($\mu = 0, 1, 2, 3$) defined by

$$x^0 \equiv ct, \quad x^1 \equiv x, \quad x^2 \equiv y, \quad x^3 \equiv z, \qquad (A.0.3)$$

and to write the invariance of the interval as

$$\eta_{\mu\nu} \, dx^\mu \, dx^\nu = \eta_{\mu\nu} \, dx^{\mu'} \, dx^{\nu'}, \qquad (A.0.4)$$

where

$$[\eta_{\mu\nu}] \equiv \begin{bmatrix} 1 & 0 & 0 & 0 \\ 0 & -1 & 0 & 0 \\ 0 & 0 & -1 & 0 \\ 0 & 0 & 0 & -1 \end{bmatrix},$$

and Einstein's summation convention has been employed (see Section 1.1). In the language of Section 2.2, we are asserting that the spacetime of special relativity is a 4-dimensional Riemannian manifold with the property that, provided *inertial* (or *cartesian*) *coordinate systems* are used, the metric tensor components $g_{\mu\nu}$ take the form $\eta_{\mu\nu}$.

Although special relativity may be formulated in arbitrary coordinate systems, we shall stick to inertial systems, and raise and lower tensor suffices using $\eta_{\mu\nu}$ or $\eta^{\mu\nu}$, where the latter are the components of the contravariant metric tensor (see Section

1.7). In terms of matrices (see remarks at end of Section 1.1), $[\eta^{\mu\nu}] = [\eta_{\mu\nu}]$, and associated tensors differ only in the signs of some of their components. For example, if $\lambda^\mu = (\lambda^0, \lambda^1, \lambda^2, \lambda^3)$, then $\lambda_\mu \equiv \eta_{\mu\nu}\lambda^\nu = (\lambda^0, -\lambda^1, -\lambda^2, -\lambda^3)$. In inertial coordinate systems, the inner product $g_{\mu\nu}\lambda^\mu\sigma^\nu = \lambda^\mu\sigma_\mu$ takes the simple form

$$\eta_{\mu\nu}\lambda^\mu\sigma^\nu = \lambda^0\sigma^0 - \lambda^1\sigma^1 - \lambda^2\sigma^2 - \lambda^3\sigma^3.$$

The frame independence contained in the second postulate is incorporated into the theory by expressing the laws of nature as tensor equations which are manifestly invariant under a change of coordinates from one inertial reference system to another.

We conclude this introduction with some remarks about time. Each inertial frame has its own coordinate time, and we shall see in the next section how these different coordinate times are related. However, it is possible to introduce an invariantly defined time associated with any given particle (or an idealised observer whose position in space may be represented by a point). The path through spacetime which represents the particle's history is called its *world line* [1], and the *proper time interval* $d\tau$ between points on its world line, whose coordinate differences relative to some frame K are dt, dx, dy, dz is defined by

$$c^2\, d\tau^2 \equiv c^2\, dt^2 - dx^2 - dy^2 - dz^2,$$

or

$$c^2\, d\tau^2 \equiv \eta_{\mu\nu}\, dx^\mu\, dx^\nu. \tag{A.0.5}$$

So

$$d\tau = (1 - v^2/c^2)^{1/2}\, dt, \tag{A.0.6}$$

where v is the particle's speed. Finite proper time intervals are obtained by integrating equation (A.0.6) along portions of the particle's world line.

Equation (A.0.6) shows that for a particle at rest in K the proper time τ is nothing other than the coordinate time t (up to an additive constant) measured by stationary clocks in K. If at any instant of the history of a moving particle we introduce an *instantaneous rest frame* K_0, such that the particle is momentarily at rest in K_0, then we see that the proper time τ is the *time recorded by a clock which moves along with the particle*. It is therefore an invariantly defined quantity, a fact which is clear from equation (A.0.5).

A.1   Lorentz transformations

A *Lorentz transformation* is a coordinate transformation connecting two inertial frames K and K'. We observed in the previous section that K' moves relative to K without rotation and with constant velocity, and it is fairly clear that this implies that the primed coordinates $x^{\mu'}$ of K' are given in terms of the unprimed coordinates x^{μ} of K via a linear (or, strictly speaking, an affine) transformation

$$x^{\mu'} = \Lambda^{\mu'}_{\nu} x^{\nu} + a^{\mu}, \tag{A.1.1}$$

where the $\Lambda^{\mu'}_{\nu}$ and a^{μ} are constants. This result also follows from the transformation formula for connection coefficients given in Exercise 2.3.1, as for a consequence of $g_{\mu\nu} = g_{\mu'\nu'} = \eta_{\mu\nu}$, $\Gamma^{\mu}_{\nu\sigma} = \Gamma^{\mu'}_{\nu'\sigma'} = 0$, and hence $X^{\mu'}_{\nu\sigma} \equiv \partial^2 x^{\mu'}/\partial x^{\nu} \, \partial x^{\sigma} = 0$, which integrates to give equation (A.1.1). Differentiation of this equation and substitution in equation (A.0.4) yields

$$\eta_{\mu\nu} = \Lambda^{\rho'}_{\mu} \Lambda^{\sigma'}_{\nu} \eta_{\rho\sigma} \tag{A.1.2}$$

as the necessary and sufficient condition for $\Lambda^{\mu'}_{\nu}$ to represent a Lorentz transformation.

If in the transformation (A.1.1) $a^{\mu} = 0$, so that the spatial origins of K and K' coincide when $t = t' = 0$, then the Lorentz transformation is called *homogeneous*, while if $a^{\mu} \neq 0$ (i.e. not all the a^{μ} are zero) then it is called *inhomogeneous*. (Inhomogeneous transformations are often referred to as *Poincaré transformations*, in which case homogeneous transformations are referred to simply as Lorentz transformations.)

To gain some insight into the meaning of a Lorentz transformation, let us consider the special case of a *boost* in the x-direction. This is the situation where the spatial origin O' of K' is moving along the x-axis of K in the positive direction with constant speed v relative to K, the axes of K and K' coinciding when $t = t' = 0$ (see Fig. A.1). The transformation is homogeneous and could take the form

$$\left.\begin{array}{l} t' = Bt + Cx, \\ x' = A(x - vt), \\ y' = y, \\ z' = z, \end{array}\right\} \tag{A.1.3}$$

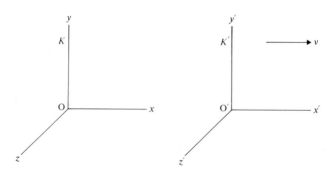

Fig. A.1   A boost in the x-direction.

the last three equations being consistent with the requirement that O' moves along the x-axis of K with speed v relative to K. Adopting this as a "trial solution" and substituting in equation (A.0.2) gives

$$B^2c^2 - A^2v^2 = c^2, \qquad BCc^2 + A^2v = 0, \qquad C^2c^2 - A^2 = -1.$$

These imply that (see Exercise A.1.1)

$$A = B = (1 - v^2/c^2)^{-1/2}, \qquad C = -(v/c^2)(1 - v^2/c^2)^{-1/2}. \tag{A.1.4}$$

If we put

$$\gamma = (1 - v^2/c^2)^{-1/2}, \tag{A.1.5}$$

then our boost may be written as

$$\left. \begin{array}{l} t' = \gamma(t - xv/c^2), \\ x' = \gamma(x - vt), \\ y' = y, \\ z' = z. \end{array} \right\} \tag{A.1.6}$$

Putting $\tanh \psi \equiv v/c$ gives (from equation (A.1.5)) $\gamma = \cosh \psi$, so the boost may also be written as

$$\left. \begin{array}{l} ct' = ct \cosh \psi - x \sinh \psi, \\ x' = x \cosh \psi - ct \sinh \psi, \\ y' = y, \\ z' = z. \end{array} \right\} \tag{A.1.7}$$

It may be shown that a general homogeneous Lorentz transformation is equivalent to a boost in some direction followed by a spatial rotation. The general inhomogeneous transformation requires an additional translation (i.e. a shift of spacetime origin).

Since $X_\nu^{\mu'} \equiv \partial x^{\mu'}/\partial x^\nu = \Lambda_\nu^{\mu'}$, a contravariant vector has components λ^μ relative to inertial frames which transform according to

$$\lambda^{\mu'} = \Lambda_\nu^{\mu'} \lambda^\nu,$$

while a covariant vector has components λ_μ which transform according to

$$\lambda_{\mu'} = \Lambda_{\mu'}^\nu \lambda_\nu,$$

where $\Lambda_{\mu'}^\nu$ is such that $\Lambda_{\mu'}^\nu \Lambda_\sigma^{\mu'} = \delta_\sigma^\nu$. These transformation rules extend to tensors. For example, a second-order mixed tensor has components τ_ν^μ which transform according to

$$\tau_{\nu'}^{\mu'} = \Lambda_\rho^{\mu'} \Lambda_{\nu'}^\sigma \tau_\sigma^\rho.$$

The equations of electromagnetism are invariant under Lorentz transformations, and in Section A.8 we present them in tensor form which brings out this invariance. However, the equations of Newtonian mechanics are not invariant under Lorentz transformations, and some modifications are necessary (see Section A.6). The transformations which leave the equations of Newtonian mechanics invariant are Galilean transformations, to which Lorentz transformations reduce when v/c is negligible (see Exercise A.1.3).

Exercises A.1

1. Verify that A, B, C are as given by equations (A.1.4).
2. What form does the matrix $[\Lambda_\nu^\mu]$ of equation (A.1.1) take for the boost of (A.1.6)? What form does its inverse $[\Lambda_{\mu'}^\nu]$ take? What is the velocity of K relative to K'?
3. Show that when v/c is negligible the equations (A.1.6) of a Lorentz boost reduce to those of a Galilean boost:

$$t' = t, \quad x' = x - vt, \quad y' = y, \quad z' = z.$$

A.2   Relativistic addition of velocities

Suppose we have three inertial frames K, K' and K'', K' being connected to K by a boost in the x-direction, and K'' being connected to K' by a boost in the x'-direction. If the speed of K' relative to K is v, then equations (A.1.7) hold, where $\tanh \psi = v/c$, and if the speed of K'' relative to K' is w, then we have analogously

$$
\left.
\begin{aligned}
ct'' &= ct' \cosh \phi - x' \sinh \phi, \\
x'' &= x' \cosh \phi - ct' \sinh \phi, \\
y'' &= y', \\
z'' &= z',
\end{aligned}
\right\} \tag{A.2.1}
$$

where $\tanh \phi = w/c$. Substituting for ct', x', y', z' from equations (A.1.7) into the above gives

$$
\left.
\begin{aligned}
ct'' &= ct \cosh (\psi + \phi) - x \sinh (\psi + \phi), \\
x'' &= x \cosh (\psi + \phi) - ct \sinh (\psi + \phi), \\
y'' &= y, \\
z'' &= z.
\end{aligned}
\right\} \tag{A.2.2}
$$

This shows that K'' is connected to K by a boost, and that K'' is moving relative to K in the positive x-direction with a speed u given by $u/c = \tanh (\psi + \phi)$. But

$$
\tanh (\psi + \phi) = \frac{\tanh \psi + \tanh \phi}{1 + \tanh \psi \tanh \phi},
$$

so

$$
u = \frac{v + w}{1 + vw/c^2}. \tag{A.2.3}
$$

This is the relativistic formula for the addition of velocities, and replaces the Newtonian formula $u = v + w$.

Note that $v < c$ and $w < c$ implies that $u < c$, so that by compounding speeds less than c one can never exceed c. For example, if $v = w = 0.75c$, then $u = 0.96c$.

Exercises A.2

1. Verify equations (A.2.2).
2. Verify that if $v < c$ and $w < c$, then the addition formula (A.2.3) implies that $u < c$.

A.3   Simultaneity

Many of the differences between Newtonian and relativistic physics are due to the concept of simultaneity. In Newtonian physics this is a frame-independent concept, whereas in relativity it is not. To see this, consider two inertial frames K and K' connected by a boost, as in Section A.1. Events which are simultaneous in K are given by $t = t_0$, where t_0 is constant. Equations (A.1.6) show that for these events

$$t' = \gamma(t_0 - xv/c^2),$$

so t' depends on x, and is not constant. The events are therefore not simultaneous in K'. (See also Fig. A.5.)

A.4   Time dilation, length contraction

Since a moving clock records its own proper time τ, equation (A.0.6) shows that the proper time interval $\Delta\tau$ recorded by a clock moving with constant speed v relative to an inertial frame K is given by

$$\Delta\tau = (1 - v^2/c^2)^{1/2} \Delta t, \qquad (A.4.1)$$

where Δt is the coordinate time interval recorded by stationary clocks in K. Hence $\Delta t > \Delta\tau$, and the moving clock "runs slow". This is the phenomenon of *time dilation*. The related phenomenon of *length contraction* (also known as *Lorentz contraction*) arises in the following way.

Suppose that we have a rod moving in the direction of its own length with constant speed v relative to an inertial frame K. There is no loss of generality in choosing this direction to be the positive x-direction of K. If K' is a frame moving in the same direction as the rod with speed v relative to K, so that K' is connected to K by a boost as in Section A.1, then the rod will be at rest in K', which is therefore a rest frame for it (see Fig. A.2). The *proper length* or *rest length* l_0 of the rod is the length as measured in the rest frame K', so

$$l_0 = x'_2 - x'_1,$$

where x'_2 and x'_1 are the x'-coordinates of its end-points in K'.

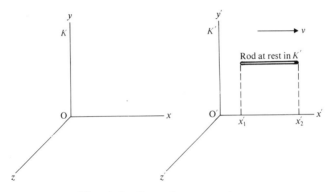

Fig. A.2   Length contraction.

According to equations (A.1.6), the x-coordinates x_1, x_2 of its end-points at any time t in K are given by

$$x'_1 = \gamma(x_1 - vt),$$
$$x'_2 = \gamma(x_2 - vt).$$

Hence if we take the difference between the end-points *at the same time t* in K, we get

$$x'_1 - x'_1 = \gamma(x_2 - x_1).$$

The length l of the rod, as measured by noting the simultaneous positions of its end-points in K, is therefore given by

$$l_0 = \gamma l, \quad \text{or} \quad l = l_0(1 - v^2/c^2)^{1/2}. \tag{A.4.2}$$

So $l < l_0$, and the moving rod is contracted.

A straightforward calculation shows that if the rod is moving relative to K in a direction perpendicular to its length, then it suffers no contraction. It follows that the volume V of a moving object, as measured by simultaneously noting the positions of its boundary points in K, is related to its reast volume V_0 by $V = V_0(1 - v^2/c^2)^{1/2}$. This fact must be taken into account when considering densities.

A.5   Spacetime diagrams

Spacetime diagrams are either three- or two-dimensional representations of spacetime, having either one or two spatial

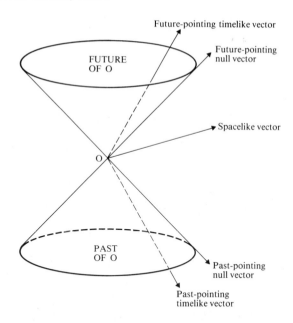

Fig. A.3   Null cone and vectors at an event O.

dimensions suppressed. When events are referred to an inertial reference system, it is conventional to orient the diagrams so that the t-axis points vertically upwards and the spatial axes are horizontal. It is also conventional to scale things so that the straight-line paths of photons are inclined at 45°; this is equivalent to using so-called relativistic units in which $c = 1$, or using the coordinates x^μ defined by equations (A.0.3).

If we consider all the photon paths passing through an event O, then these constitute the *null cone* at O (see Fig. A.3). The region of spacetime contained within the upper half of the null cone is the *future* of O, while that contained within the lower half is its *past*. The region outside the null cone contains events which may either come before or after the event O in time, depending on the reference system used, but there is no such ambiguity about the events in the future and in the past. This follows from the fact that the null cone at O is invariantly defined. If the event O is taken as the origin of an inertial reference system, then the equation of the null cone is

$$x^2 + y^2 + z^2 = c^2 t^2. \tag{A.5.1}$$

If we have a vector λ^μ localised at O, then λ^μ is called *timelike* if it lies within the null cone, *null* if it is tangential to the null cone, and *spacelike* if it lies outside the null cone. That is, λ^μ is

$$
\begin{array}{lll}
\text{timelike} & & \left. \begin{array}{l} > 0 \end{array} \right. \\
\text{null} & \text{if} \quad \eta_{\mu\nu}\lambda^\mu\lambda^\nu & \left\{ \begin{array}{l} = 0. \\ \end{array} \right. \\
\text{spacelike} & & \left. \begin{array}{l} < 0 \end{array} \right.
\end{array} \qquad \text{(A.5.2)}
$$

Timelike and null vectors may be characterised further as *future-pointing* or *past-pointing* (see Fig. A.3).

Consider now the world line of a particle with mass. Relativistic mechanics prohibit the acceleration of such a particle to speeds up to c (a fact suggested by the formula (A.2.3) for the addition of velocities) [2], which implies that its world line must lie within the null cone at each event on it, as the following remarks show. With the speed $v < c$, the proper time τ as defined by equation (A.0.6) is real, and may be used to parametrise the world line: $x^\mu = x^\mu(\tau)$. Its tangent vector $u^\mu \equiv dx^\mu/d\tau$ (see Section 2.2) is called the *world velocity* of the particle, and equation (A.0.5) shows that

$$\eta_{\mu\nu}u^\mu u^\nu = c^2,$$

so u^μ is timelike and lies within the null cone at each event on the world line (see Fig. A.4(a)). The tangent vector at each event on the world line of a photon is clearly null (see Fig. A.4(b)).

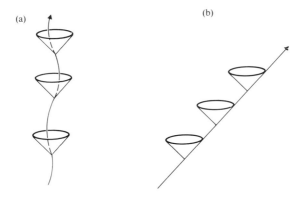

(a)   (b)

Fig. A.4   World lines of (a) a particle with mass, and (b) a photon.

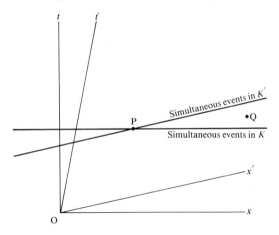

Fig. A.5   Spacetime diagram of a boost.

Spacetime diagrams may be used to illustrate Lorentz transformations. A 2-dimensional diagram suffices to illustrate the boost of Section A.1 connecting the frames K and K'. The x'-axis of K' is given by $t' = 0$, i.e. by $t = xv/c^2$, while the t'-axis of K' is given by $x' = 0$, i.e. by $x = vt$. So with $c = 1$, the slope of the x'-axis relative to K is v, while that of the t'-axis is $1/v$. So if the axes of K are drawn orthogonal as in Fig. A.5, then those of K' are not orthogonal, but inclined as shown.

Events which are simultaneous in K are represented by a line parallel to the x-axis, while those which are simultaneous in K' are represented by a line parallel to the x'-axis, and the frame-dependence of the concept of simultaneity is clearly illustrated in a spacetime diagram. Note that the event Q of Figure A.5 occurs *after* the event P according to observers in K, while it occurs *before* the event P according to observers in K'.

Exercises A.5

1. Check the criterion (A.5.2).

A.6   Some standard 4-vectors

Here we introduce some standard 4-vectors of special relativity, and comment briefly on their roles in relativistic mechanics. The prefix 4- serves to distinguish vectors in spacetime from those in space, which we shall call 3-vectors. It is useful to introduce the notation

$$\lambda^\mu = (\lambda^0, \lambda^1, \lambda^2, \lambda^3) = (\lambda^0, \boldsymbol{\lambda}), \quad (A.6.1)$$

so that bold-faced letters represent spatial parts.

We have already defined the world velocity $u^\mu \equiv dx^\mu/d\tau$ of a particle with mass. If we introduce the *coordinate velocity* v^μ (which is not a 4-vector) defined by

$$v^\mu \equiv dx^\mu/dt = (c, \mathbf{v}), \quad (A.6.2)$$

where $\mathbf{v}$ is the particle's 3-velocity, then

$$u^\mu = (dt/d\tau)v^\mu = (\gamma c, \gamma \mathbf{v}), \quad (A.6.3)$$

where $\gamma = (1 - v^2/c^2)^{-1/2}$. The particle's 4-*momentum* p^μ is defined in terms of u^μ by

$$p^\mu \equiv mu^\mu, \quad (A.6.4)$$

where m is the particle's rest mass [3]. The zeroth component p^0 is E/c, where E is the energy of the particle, and we can put

$$p^\mu = (E/c, \mathbf{p}). \quad (A.6.5)$$

The wave aspect of light may be built into the particle approach of photons by associating with a photon a *wave* 4-*vector* k^μ defined by

$$k^\mu \equiv (2\pi/\lambda, \mathbf{k}), \quad (A.6.6)$$

where λ is the wavelength and $\mathbf{k} = (2\pi/\lambda)\mathbf{n}$, $\mathbf{n}$ being a unit 3-vector in the direction of propagation [4]. It follows that $k^\mu k_\mu = 0$, so that k^μ is null. It is, of course, tangential to the photon's world line. The photon's 4-momentum p^μ is given by

$$p^\mu \equiv (h/2\pi)k^\mu, \quad (A.6.7)$$

where h is Planck's constant. Thus the photon's energy is

$$E = cp^0 = hc/\lambda = h\nu,$$

where ν is the frequency, in agreement with the quantum-mechanical result.

In relativistic mechanics, Newton's second law is modified to

$$dp^\mu/d\tau = f^\mu, \tag{A.6.8}$$

where f^μ is the 4-*force* on the particle. This is given in terms of the 3-force $\mathbf{F}$ by

$$f^\mu \equiv \gamma(\mathbf{F} \cdot \mathbf{v}/c, \mathbf{F}). \tag{A.6.9}$$

The invariance of the inner product gives

$$p^\mu p_\mu = p^{\mu'} p_{\mu'}. \tag{A.6.10}$$

If we take the primed frame K' to be an instantaneous rest frame, then $p^{\mu'} = (mc, \mathbf{0})$, and the right-hand side is $m^2 c^2$. The left-hand side is $E^2/c^2 - \mathbf{p} \cdot \mathbf{p}$, so equation (A.6.10) gives

$$E = (p^2 c^2 + m^2 c^4)^{1/2}, \tag{A.6.11}$$

where $p^2 = \mathbf{p} \cdot \mathbf{p}$. This is the well-known result connecting the energy E of a particle with its momentum and rest mass.

From equations (A.6.3) and (A.6.5) we see that

$$\mathbf{p} = \gamma m \mathbf{v} \tag{A.6.12}$$

and that $E/c = p^0 = \gamma mc$, so

$$E = \gamma mc^2 = mc^2(1 - v^2/c^2)^{-1/2}$$
$$= mc^2 + \tfrac{1}{2}mv^2 + \dots . \tag{A.6.13}$$

Equation (A.6.12) shows that the spatial part $\mathbf{p}$ of the relativistic 4-momentum p^μ reduces to the Newtonian 3-momentum $m\mathbf{v}$ when v is small compared to c (giving $\gamma \simeq 1$). However, equation (A.6.13) shows that E reduces to $mc^2 + \tfrac{1}{2}mv^2$, and that the total energy includes not only the kinetic energy $\tfrac{1}{2}mv^2$, but also the *rest energy* mc^2, the latter being unsuspected in Newtonian physics. It should be noted that we have *not proved* the celebrated formula $E = \gamma mc^2$; it simply follows from our defining E by $p^0 = E/c$. Although this definition is standard in relativity, it is sensible to ask how it ever came about.

Conservation of momentum is an extremely useful principle, and if we wish to preserve it in relativity, then it turns out that we must define momentum $\mathbf{p}$ by $\mathbf{p} = \gamma m \mathbf{v}$ rather than $\mathbf{p} = m\mathbf{v}$. This follows from a consideration of simple collision problems in different inertial frames [5]. But $\gamma m \mathbf{v}$ is the spatial part of the 4-vector p^μ defined by equation (A.6.4), and it follows from equations (A.6.8) and (A.6.9) that $\mathrm{d}p^0/\mathrm{d}\tau = (\gamma/c)\mathbf{F} \cdot \mathbf{v}$, which implies that

$$\mathbf{F} \cdot \mathbf{v} = c \, \mathrm{d}p^0/\mathrm{d}t.$$

But $\mathbf{F} \cdot \mathbf{v}$ is the rate at which the 3-force $\mathbf{F}$ imparts energy to the particle, hence it is natural to define the energy E of the particle by $E = cp^0$. The conservation of energy and momentum of a free particle is then incorporated in the single 4-vector equation

$$p^\mu = \text{constant.}$$

This extends to a system of interacting particles with no external forces:

$$\sum_{\substack{\text{all} \\ \text{particles}}} p^\mu = \text{constant.}$$

As an illustration of this, consider the Compton effect in which a photon collides with a stationary electron (see Fig. A.6). Initially the photon is travelling along the x^1-axis of our reference system and it collides with an electron at rest. After collision the electron and photon move off in the plane $x^3 = 0$, making angles θ and ϕ with the x^1-axis as shown. Remembering that the energy

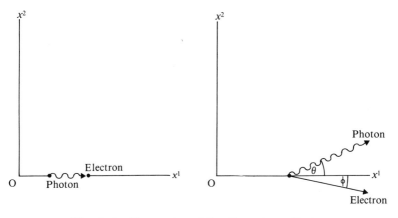

Fig. A.6 Geometry of the Compton effect.

of a photon is $h\nu$, and that for a photon $p^\mu p_\mu = 0$, we have before collision:

$$p^\mu_{\text{ph}} = (h\nu/c, h\nu/c, 0, 0),$$

$$p^\mu_{\text{el}} = (mc, 0, 0, 0),$$

and after collision

$$\bar{p}^\mu_{\text{ph}} = (h\bar{\nu}/c, (h\bar{\nu}/c)\cos\theta, (h\bar{\nu}/c)\sin\theta, 0),$$

$$\bar{p}^\mu_{\text{el}} = (\gamma mc, \gamma mv\cos\phi, -\gamma mv\sin\phi, 0),$$

where v is the electron's speed after collision. The conservation laws contained in

$$p^\mu_{\text{ph}} + p^\mu_{\text{el}} = \bar{p}^\mu_{\text{ph}} + \bar{p}^\mu_{\text{el}}$$

then give

$$h\nu/c + mc = h\bar{\nu}/c + \gamma mc,$$

$$h\nu/c = (h\bar{\nu}/c)\cos\theta + \gamma mv\cos\phi,$$

$$0 = (h\bar{\nu}/c)\sin\theta - \gamma mv\sin\phi.$$

Eliminating v and ϕ from these leads to the formula for Compton scattering (see Exercise A.6.2) giving the frequency of the photon after collision as

$$\bar{\nu} = \frac{\nu}{1 + (h\nu/mc^2)(1 - \cos\phi)}. \tag{A.6.14}$$

Exercises A.6

1. Show that as a consequence of equation (A.6.8) we have $u_\mu f^\mu = 0$, and that f^μ as given by equation (A.6.9) satisfies this relation.
2. Check the derivation of the formula (A.6.14).

A.7   Doppler effect

Suppose that we have a source of radiation which is moving relative to an inertial frame K with speed v in the positive x^1-direction in the plane $x^3 = 0$, and that at some instant an observer fixed at the origin O of K receives a photon in a direction which makes an angle θ with the positive x^1-direction (see Fig. A.7).

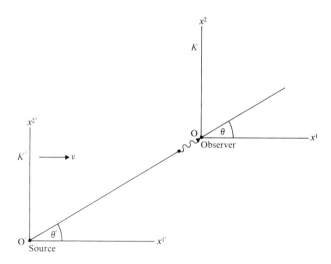

Fig. A.7   Photon arriving at an observer from a moving source.

Let us attach to the source a frame K' whose axes are parallel to those of K, and which moves along with the source, so that it is at rest in K' at the origin O'. The frame K' is therefore connected to K by a Lorentz transformation comprising a boost in the x^1-direction and a translation (see Section A.1). It follows that the wave 4-vector k^μ of the photon transforms according to

$$k^{\mu'} = \Lambda^{\mu'}_\nu k^\nu, \tag{A.7.1}$$

where

$$[\Lambda^{\mu'}_\nu] = \begin{bmatrix} \gamma & -\gamma v/c & 0 & 0 \\ -\gamma v/c & \gamma & 0 & 0 \\ 0 & 0 & 1 & 0 \\ 0 & 0 & 0 & 1 \end{bmatrix}$$

(see Exercise A.1.2). The zeroth component gives $k^{0'} = \Lambda^{0'}_\nu k^\nu$, or

$$k^{0'} = \gamma k^0 - (\gamma v/c)k^1. \tag{A.7.2}$$

Now $k^{0'} = 2\pi/\lambda_0$, where λ_0 is the *proper wavelength* as observed in the frame K' in which the source is at rest, and

$$k^\mu = (2\pi/\lambda)(1, \cos\theta, \sin\theta, 0),$$

where λ is the wavelength as observed by the observer at the origin O of K. (Note that we are making use of the fact that k^μ

is constant along the photon's world line.) Hence equation (A.7.2) gives

$$\frac{1}{\lambda_0} = \frac{1}{\lambda}\left(\gamma - \frac{\gamma v}{c}\cos\theta\right),$$

so

$$\lambda/\lambda_0 = \gamma[1 - (v/c)\cos\theta], \tag{A.7.3}$$

and the observed wavelength is different from the proper wavelength.

If the source is on the negative x^1-axis, so that it is approaching the observer, then $\theta = 0$ and equation (A.7.3) gives

$$\frac{\lambda}{\lambda_0} = \gamma(1 - v/c) = \left(\frac{1 - v/c}{1 + v/c}\right)^{1/2}. \tag{A.7.4}$$

Thus $\lambda < \lambda_0$ and the observed wavelength is blue-shifted.

If the source is on the positive x^1-axis, so that it is receding from the observer, then $\theta = \pi$ and equation (A.7.3) gives

$$\frac{\lambda}{\lambda_0} = \gamma(1 + v/c) = \left(\frac{1 + v/c}{1 - v/c}\right)^{1/2}. \tag{A.7.5}$$

In this case $\lambda > \lambda_0$, and the observed wavelength is red-shifted.

If the source is displaced away from the x^1-axis, then at some instant we will have $\theta = \pm\pi/2$ giving

$$\lambda/\lambda_0 = \gamma = (1 - v^2/c^2)^{-1/2}. \tag{A.7.6}$$

This shifting of the observed spectrum is known as the *Doppler effect*. The formulae (A.7.4) and (A.7.5) refer to approach and recession, and have their counterparts in non-relativistic physics. The formula (A.7.6) is that of the *transverse* Doppler effect, and has no such counterpart. The transverse effect was observed in 1938 by Ives and Stillwell, who examined the spectra of rapidly moving hydrogen atoms. The formula (A.7.6) may be used in discussions of the celebrated twin paradox [6].

Exercises A.7

1. Using equation (A.7.1), show that the angle θ' (see Fig. A.7) at which the photon leaves the source, as measured in K', is

given by

$$\tan \theta' = \frac{\tan \theta}{\gamma[1 - (v/c) \sec \theta]}.$$

(This is essentially the *relativistic aberration formula*.)

A.8   Electromagnetism

The equations which govern the behaviour of the electromagnetic field in free space are Maxwell's equations, which in SI units take the form

$$\nabla \cdot \mathbf{B} = 0, \tag{A.8.1a}$$

$$\nabla \cdot \mathbf{E} = \rho/\varepsilon_0, \tag{A.8.1b}$$

$$\nabla \times \mathbf{E} = -\partial \mathbf{B}/\partial t, \tag{A.8.1c}$$

$$\nabla \times \mathbf{B} = \mu_0 \mathbf{J} + \mu_0 \varepsilon_0 \, \partial \mathbf{E}/\partial t. \tag{A.8.1d}$$

Here $\mathbf{E}$ is the *electric field intensity*, $\mathbf{B}$ is the *magnetic induction*, ρ is the charge density (charge per unit volume), $\mathbf{J}$ is the current density, μ_0 is the *permeability* of free space, and ε_0 is the *permittivity* of free space. The last two quantities satisfy

$$\mu_0 \varepsilon_0 = 1/c^2. \tag{A.8.2}$$

The vector fields $\mathbf{B}$ and $\mathbf{E}$ may be expressed in terms of a *vector potential* $\mathbf{A}$ and a *scalar potential* ϕ:

$$\mathbf{B} = \nabla \times \mathbf{A}, \qquad \mathbf{E} = -\nabla \phi - \partial \mathbf{A}/\partial t. \tag{A.8.3}$$

Equations (A.8.1a) and (A.8.1c) are then satisfied. These potentials are not uniquely determined by Maxwell's equations, and $\mathbf{A}$ may be replaced by $\mathbf{A} + \nabla \psi$ and ϕ by $\phi - \partial \psi/\partial t$, where ψ is arbitrary. Such transformations of the potentials are known as *gauge transformations*, and allow one to choose $\mathbf{A}$ and ϕ so that they satisfy the *Lorentz gauge condition*, which is

$$\nabla \cdot \mathbf{A} + \varepsilon_0 \mu_0 \, \partial \phi/\partial t = 0. \tag{A.8.4}$$

The remaining two Maxwell equations (A.8.1b) and (A.8.1d) then imply that $\mathbf{A}$ and ϕ satisfy

$$\Box \mathbf{A} = -\mu_0 \mathbf{J}, \qquad \Box \phi = -\rho/\varepsilon_0, \tag{A.8.5}$$

where $\square$ is the *d'Alembertian* defined by

$$\square \equiv \nabla^2 - c^{-2} \, \partial^2/\partial t^2. \qquad (A.8.6)$$

Equations (A.8.5) may be solved in terms of retarded potentials, and the form of the solution shows that we may take ϕ/c and $\mathbf{A}$ as the temporal and spatial parts of a 4-vector

$$A^\mu \equiv (\phi/c, \mathbf{A}), \qquad (A.8.7)$$

which is known as the *4-potential* [7].

Maxwell's equations take on a particularly simple and elegant form if we introduce the *electromagnetic field tensor* $F_{\mu\nu}$ defined by

$$F_{\mu\nu} \equiv A_{\mu,\nu} - A_{\nu,\mu}, \qquad (A.8.8)$$

where $A_\mu = (\phi/c, -\mathbf{A})$ is the covariant 4-potential, and commas denote partial derivatives. Equations (A.8.3) show that

$$[F_{\mu\nu}] = \begin{bmatrix} 0 & -E^1/c & -E^2/c & -E^3/c \\ E^1/c & 0 & B^3 & -B^2 \\ E^2/c & -B^3 & 0 & B^1 \\ E^3/c & B^2 & -B^1 & 0 \end{bmatrix}, \qquad (A.8.9)$$

where $(E^1, E^2, E^3) \equiv \mathbf{E}$ and $(B^1, B^2, B^3 \equiv \mathbf{B}$. It is then a straightforward process (using the result of Exercise A.8.3) to check that Maxwell's equations are equivalent to

$$F^{\mu\nu}{}_{,\nu} = \mu_0 j^\mu, \qquad (A.8.10a)$$

$$F_{\mu\nu,\sigma} + F_{\nu\sigma,\mu} + F_{\sigma\mu,\nu} = 0, \qquad (A.8.10b)$$

where $j^\mu \equiv (\rho c, \mathbf{J})$ is the 4-*current density*. Note that $j^\mu = \rho v^\mu = \rho_0 u^\mu$, where u^μ is the world velocity of the charged particles producing the current distribution, and ρ_0 is the *proper* charge density. That is, ρ_0 is the charge per unit *rest* volume, whereas ρ is the charge per unit volume (see remark at end of Section A.4).

The equation of motion of a particle of charge q moving in an electromagnetic field is

$$d\mathbf{p}/dt = q(\mathbf{E} + \mathbf{v} \times \mathbf{B}), \qquad (A.8.11)$$

where $\mathbf{p}$ is its momentum and $\mathbf{v}$ its velocity. The right-hand side of this equation is known as the *Lorentz force*. It follows that the rate at which the electromagnetic field imparts energy E to the particle is given by

$$dE/dt = \mathbf{F} \cdot \mathbf{v} = q\mathbf{E} \cdot \mathbf{v}. \qquad (A.8.12)$$

Equations (A.8.11) and (A.8.12) may be brought together in a single 4-vector equation (see Exercise A.8.5):

$$dp^\mu/d\tau = -qF^\mu{}_\nu u^\nu, \qquad (A.8.13)$$

or

$$m\, d^2x^\mu/d\tau^2 = -qF^\mu{}_\nu u^\nu, \qquad (A.8.14)$$

where m is the rest mass of the particle. The continuous version of equation (A.8.14) is

$$\mu\, d^2x^\mu/d\tau^2 = -F^\mu{}_\nu j^\nu, \qquad (A.8.15)$$

where μ is the proper (mass) density of the charge distribution giving rise to the electromagnetic field, and this is the equation of motion of a charged unstressed fluid. That is, the only forces acting on the fluid particles arise from their electromagnetic interaction, there being no body forces nor mechanical stress forces such as pressure.

It is evident that Maxwell's equations and related equations may be formulated as 4-vector and tensor equations without modification. They are therefore invariant under Lorentz transformations, but not under Galilean transformations, and this observation played a leading role in the development of special relativity. By contrast, the equations of Newtonian mechanics are invariant under Galilean transformations, but not under Lorentz transformations, and therefore require modification to incorporate them into special relativity.

Exercises A.8

1. Show that the Lorentz gauge condition (A.8.4) may be written as $A^\mu{}_{,\mu} = 0$.

2. Check that the components $F_{\mu\nu}$ are as displayed in equation (A.8.9).

3. Show that the mixed and contravariant forms of the electromagnetic field tensor are given by

$$[F^{\mu}{}_{\nu}] = \begin{bmatrix} 0 & -E^1/c & -E^2/c & -E^3/c \\ -E^1/c & 0 & -B^3 & B^2 \\ -E^2/c & B^3 & 0 & -B^1 \\ -E^3/c & -B^2 & B^1 & 0 \end{bmatrix},$$

$$[F^{\mu\nu}] = \begin{bmatrix} 0 & E^1/c & E^2/c & E^3/c \\ -E^1/c & 0 & B^3 & -B^2 \\ -E^2/c & -B^3 & 0 & B^1 \\ -E^3/c & B^2 & -B^1 & 0 \end{bmatrix}.$$

4. Verify that equations (A.8.10) are equivalent to Maxwell's equations.

5. Verify that equations (A.8.11) and (A.8.12) may be brought together in the single 4-vector equation (A.8.13).

Problems A

1. If an astronaut claims that a spaceflight took 3 days, while a base station on Earth claims that it took 3.000 001 5 days, what kind of average rocket speed are we talking about?

2. Illustrate the phenomena of time dilation and length contraction using spacetime diagrams.

3. If a laser in the laboratory has a wavelength of 632.8 nm, what wavelength would be observed by an observer approaching it directly at a speed $c/2$?

4. Show that under a boost in the x^1-direction the components of the electric field intensity $\mathbf{E}$ and the magnetic induction $\mathbf{B}$ transform according to

$$E^{1'} = E^1, \qquad\qquad B^{1'} = B^1,$$
$$E^{2'} = \gamma(E^2 - vB^2), \qquad B^{2'} = \gamma(B^2 + vE^3/c^2),$$
$$E^{3'} = \gamma(E^3 + vB^2), \qquad B^{3'} = \gamma(B^3 - vE^2/c^2).$$

5. Plot a graph of $\gamma(v) \equiv (1 - v^2/c^2)^{-1/2}$.

6. In a laboratory a certain switch is turned on, and then turned

off 3 seconds later. In a "rocket frame" these events are found to be separated by 5 seconds. Show that in the rocket frame the spatial separation between the two events is 12×10^8 m, and that the rocket frame has a speed 2.4×10^8 m s^{-1} (i.e. $4c/5$) relative to the laboratory. (Take $c = 3 \times 10^8$ m s^{-1}.)

7. A uniform charge distribution of proper density ρ_0 is at rest in an inertial frame K. Show that an observer moving with a velocity $\mathbf{v}$ relative to K sees a charge density $\gamma\rho_0$ and a current density $-\gamma\rho_0\mathbf{v}$.

8. Show that the Doppler-shift formula (A.7.3) may be expressed invariantly as

$$\lambda/\lambda_0 = (u_0^\mu k_\mu)/(u^\nu k_\nu),$$

where λ is the observed wavelength, λ_0 the proper wavelength, k^μ the wave 4-vector, u^μ the world velocity of the observer and u_0^μ that of the source.

9. It is found that a stationary "cupful" of radioactive pions have a half-life of 1.77×10^{-8} seconds. A collimated pion beam leaves an accelerator at a speed of $0.99c$, and it is found to drop to half its original intensity 37.3 m away. Are these results consistent?

(Look at the problem from two separate viewpoints, namely that of time dilation and that of path contraction. Take $c = 3 \times 10^8$ m s^{-1}.)

10. Caesium-beam clocks have been taken at high speeds around the world in commercial jets. Show that for an equatorial circumnavigation at a height of 9 km (about 30 000 feet) and a speed of 250 m s^{-1} (about 600 m.p.h.) one would expect, *on the basis of special relativity alone*, the following time gains (or losses), when compared with a clock which remains fixed on Earth:

Westbound flight	*Eastbound flight*
$+150 \times 10^{-9}$ seconds	-262×10^{-9} seconds

(Begin by considering why there is a difference for westbound and eastbound flights. Take $R_\oplus = 6378$ km, $c = 3 \times 10^8$ m s^{-1}. In 1971 Hafele and Keating [8] performed experiments along these lines, primarily to check the effect that the Earth's gravitational field had on the rates of clocks, which is to be ignored in this calculation.)

Notes

1. For an extended object we have a *world tube*.
2. Particles having speeds in excess of c, called tachyons, have been postulated, but attempts to detect them have (to date) been unsuccessful. They cannot be decelerated to speeds below c.
3. As in Chapter 3, we use m rather than the more usual notation m_0 for rest mass.
4. The factor 2π, which seems to be an encumbrance, simplifies expressions in relativistic optics and wave theory.
5. See, for example, Rindler, 1960, §28. Note that in Rindler m_0 is proper mass and m is *relativistic mass*; in our notation these quantities are m and γm respectively.
6. See Feenberg, 1959.
7. See, for example, Rindler, 1960, §41.
8. Hafele and Keating, 1972.

References

Anderson, J. D., Esposito, P. B., Martin, W. and Thornton, C. L. (1975) 'Experimental test of general relativity using time-delay data from *Mariner 6* and *Mariner 7*', *Astrophys. J.*, **200**, 221–33.

Apostol, T. M. (1974) *Mathematical Analysis*, 2nd edn, Addison-Wesley, Reading, Mass.

Birkhoff, G. and MacLane, S. (1977) *A Survey of Modern Algebra*, 4th edn, Macmillan, New York (Collier-Macmillan, London).

Bondi, H. (1960) *Cosmology*, 2nd edn, Cambridge University Press, Cambridge.

Carmeli, M., Fickler, S. I. and Witten, L., eds (1970) *Relativity*, Plenum, New York.

Charlier, C. V. I. (1922) 'How an infinite world may be built up', *Ark. Mat. Astron. Fys.*, **16**, No. 22.

Duncombe, R. L. (1956) 'Relativity effects for the three inner planets', *Astronom. J.*, **61**, 174–5.

Feenberg, E. (1959) 'Doppler effect and time dilatation', *Am. J. Phys.*, **27**, 190.

Friedmann, A. (1922) 'Über die Krümmung des Raumes', *Z. Phys.*, **10**, 377–86.

Goldstein, H. (1959) *Classical Mechanics*, Addison-Wesley, Reading, Mass.

Hafele, J. C. and Keating, R. E. (1972) 'Around-the-world atomic clocks: observed relativistic time gains', *Science*, **177**, 168–70.

Halmos, P. R. (1974) *Finite-Dimensional Vector Spaces*, 2nd edn, Springer, New York and Berlin (originally published by Van Nostrand).

Hawking, S. W. and Ellis, G. F. R. (1973) *The Large Scale Structure of Space-Time*, Cambridge University Press, Cambridge.

Hoffman, B. (1972) *Albert Einstein: Creator and Rebel*, Hart-Davies, MacGibbon, London (New American Library, New York).

Kilmister, C. W. (1973) *General Theory of Relativity*, Pergamon, Oxford.

Landau, L. D. and Lifshitz, E. M. (1959) *Fluid Mechanics*, Pergamon, Oxford.

Landau, L. D. and Lifshitz, E. M. (1975) *The Classical Theory of Fields*, 4th edn, Pergamon, Oxford.

Misner, C. W., Thorne, K. S. and Wheeler, J. A. (1973) *Gravitation*, Freeman, San Francisco.

Møller, C. (1972) *The Theory of Relativity*, 2nd edn, Oxford University Press, Oxford.

Penzias, A. A. and Wilson, R. W. (1965) 'A measurement of excess antenna temperature at 4080 Mc/s', *Astrophys. J.*, **142**, 419–21.

Pound, R. V. and Rebka, G. A. (1960) 'Apparent weight of photons', *Phys. Rev. Lett.*, **4**, 337–41.

Riley, J. M. (1973) 'A measurement of the gravitational deflection of radio waves by the Sun during 1972 October', *Mon. Not. R. Astr. Soc.*, **161**, 11P–14P.

Rindler, W. (1960) *Special Relativity*, Oliver and Boyd, Edinburgh.

Robertson, H. P. (1935) 'Kinematics and world structure', *Astrophys. J.*, **82**, 248–301.

Robertson, H. P. (1936) 'Kinematics and world structure', *Astrophys. J.*, **83**, 257–71.

Schouten, J. A. (1954) *Ricci-Calculus*, 2nd edn, Springer, Berlin.

Shapiro, I. I. (1964) 'Fourth test of general relativity', *Phys. Rev. Lett.*, **13**, 789–91.

Shapiro, I. I. (1968) 'Fourth test of general relativity: preliminary results', *Phys. Rev. Lett.*, **20**, 1265–9.

Shapiro, I. I., Ash, M. E., Campbell, D. B., Dyce, R. B., Ingalls, R. P., Jurgens, R. F. and Pettingill, G. H. (1971) 'Fourth test of general relativity: new radar results', *Phys. Rev. Lett.*, **26**, 1132-5.

Thorne, K. S. (1974) 'The search for black holes', *Scientific American*, **231**, No. 6, 32–43.

Walker, A. G. (1936) 'On Milne's theory of world structure', *Proc. London Math. Soc.*, **42**, 90–127.

Weinberg, S. (1972) *Gravitation and Cosmology: Principles and Applications of the General Theory of Relativity*, Wiley, New York.

Young, P. J., Westphal, J. A., Kristian, J., Wilson, C. P. and Landauer, F. P. (1978) 'Evidence for a supermassive object in the nucleus of the galaxy M87 from SIT and CCD area photometry', *Astrophys. J.* **221**, 721–30.

Index